T0342395

**Interval Methods for Uncertain Power
System Analysis**

Interval Methods for Uncertain Power System Analysis

Alfredo Vaccaro
University of Sannio
Italy

IEEE Press Series on Power and Energy Systems
Ganesh Kumar Venayagamoorthy, Series Editor

IEEE PRESS
WILEY

Published by John Wiley & Sons, Inc., Hoboken, New Jersey.
Published simultaneously in Canada.

For general information on our other products and services or for technical support, please
contact our Customer Care Department within the United States at (800) 762-2974, outside the
United States at (317) 572-3993 or fax (317) 572-4002.

Wiley also publishes its books in a variety of electronic formats. Some content that appears in
print may not be available in electronic formats. For more information about Wiley products,
visit our web site at www.wiley.com.

Library of Congress Cataloging-in-Publication Data:

Names: Vaccaro, Alfredo, author. | John Wiley & Sons, publisher.
Title: Interval methods for uncertain power system analysis / Alfredo
    Vaccaro.
Description: Hoboken, New Jersey : Wiley-IEEE Press, [2023] | Includes
    index.
Identifiers: LCCN 2023014418 (print) | LCCN 2023014419 (ebook) | ISBN
    9781119855040 (cloth) | ISBN 9781119855057 (adobe pdf) | ISBN
    9781119855064 (epub)
Subjects: LCSH: Electric power systems. | Electric power
    systems–Mathematical models. | Electric power systems–Reliability. |
    System analysis.
Classification: LCC TK1005 .V25 2023 (print) | LCC TK1005 (ebook) | DDC
    621.3101/5118–dc23/eng/20230506
LC record available at https://lccn.loc.gov/2023014418
LC ebook record available at https://lccn.loc.gov/2023014419

Cover Design: Wiley
Cover Image: © Sukpaiboonwat/Shutterstock

Set in 9.5/12.5pt STIXTwoText by Straive, Chennai, India

To my family

Contents

About the Author

Alfredo Vaccaro, PhD, is a full professor of electric power systems at the Department of Engineering of University of Sannio. He is the editor-in-chief of Smart Grids and Sustainable Energy, Springer Nature and associate editor of IEEE trans. on Smart Grids, IEEE trans. on Power Systems, and IEEE Power Engineering Letters.

Preface

This book summarizes the main results of my research activities in the field of uncertain power system analysis by interval methods.

I started working on this interesting and challenging issue in early 2000, inspired by the papers of Prof. Fernando Alvarado about the application of interval arithmetic in uncertain power flow analysis. The first contributions were focused on mitigating the effects of the "dependency problem" and "wrapping effect" in interval analysis, which can reduce the value of the results by overestimating the bounds of the power flow solutions. This overestimation problem has been observed when solving many conventional power system operation problems by "naive" interval analysis, which may lead to aberrant solutions, due to the inability of interval arithmetic to model the correlations between the uncertain variables. Consequently, each step of the algorithm introduces spurious values, causing the solution bounds to converge to overconservative values. This phenomena has been extensively studied in qualitative systems analysis and requires the use of complex and time-consuming preconditioning techniques.

This stimulated the research into alternative interval methods based on Affine Arithmetic, which is one of the main topics of this book.

In this approach, each uncertain variable is described by a first-degree polynomial composed of a central value and a number of partial deviations, each one modeling the effect of an independent source of uncertainty.

The adoption of Affine Arithmetic makes it possible to solve uncertain mathematical programming problems and obtain a reliable estimation of the solution hulls by including the effect of correlation between the uncertain variables, as well as the heterogeneity of the sources of uncertainty.

This reliable computing paradigm has been successfully deployed to solve a large number of power system operation problems in the presence of data uncertainty. The examples presented in this book cover power flow studies, optimal power flow analysis, dynamic thermal rating assessment, state estimation, and stability analysis.

The obtained results demonstrate the effectiveness of interval methods to solve the complex uncertain problems encountered when analyzing realistic power system operation scenarios, hence, making these methods one of the most promising alternatives for stochastic information management in modern power systems.

May 2023

Alfredo Vaccaro
Benevento, Italy

Acknowledgments

I wish to express my sincere gratitude to my mentor Prof. Claudio Canizares, who inspired and stimulated my research activities in the field of uncertain power system analysis.

I would also like to thank Prof. Kankar Bhattacharya for his valuable and qualified suggestions about the potential role of range analysis in market studies, and Dr. Adam J. Collin and Dr. Fabrizio De Caro for their valuable support in reviewing this book.

Alfredo Vaccaro

Acronyms

AA	affine arithmetic
DTR	dynamic thermal rating
IA	interval analysis
IM	interval mathematics
MC	Markov Chain
OPF	optimal power flow
PF	power flow

Introduction

Power system analysis is often affected by large and correlated uncertainties, which could seriously affect the validity of the obtained results. Uncertainties in modern power systems stem from both internal and external factors, including model inaccuracies, measurement errors, inconsistent data, and imprecise knowledge about some input information.

Conventional methods for uncertainty modeling in power system analysis involve using probabilistic techniques to characterize the variability in input data and sampling-based approaches to simulate the system behavior for a large number of possible operating scenarios.

However, relying on probabilistic methods has its limitations, as power system engineers may struggle to express their imprecise knowledge about certain input variables with probability distributions (due to the subjective and qualitative nature of their expertise), and, more generally, there may be a lack of reliable data for characterizing the probability parameters. Additionally, the use of probabilistic methods often requires the assumptions of normal distributions and statistical independence, e.g. in the case of weather variables, but operational experience shows that these assumptions are frequently unsupported by empirical evidence.

Recent advances in the field have expanded the range of methods for addressing uncertainty by introducing a number of nonprobabilistic techniques, such as interval analysis, fuzzy arithmetic, and evidence theory.

Nonprobabilistic techniques are often used when uncertainty arises from limitations in our understanding of the system, rather than unpredictable numerical data. In these cases, only rough estimates of values and relationships between variables are available. Consider, for example the power profiles generated by small and dispersed renewable generators, which strictly depend on the evolution of some weather variables, e.g. solar radiation for photovoltaic generators and wind speed and direction for wind turbines. Although these variables can be measured at a specific location, it is challenging to determine their distributions over a wide geographical area. Nonetheless, weather forecasts periodically provide

Interval Methods for Uncertain Power System Analysis, First Edition. Alfredo Vaccaro.
© 2023 The Institute of Electrical and Electronics Engineers, Inc. Published 2023 by John Wiley & Sons, Inc.

qualitative information about the expected evolution of the weather variables on different time horizons, but these predicted profiles cannot be easily expressed as probabilities.

It follows that the availability of reliable frameworks for modeling and managing nonprobabilistic knowledge can greatly enhance the robustness and effectiveness of power system analysis.

For this purpose, interval methods have been recognized as an enabling methodology for uncertain power system analysis.

The primary benefit of these techniques is that they intrinsically keep track of the computing accuracy of each elementary mathematical operation, without needing information or assumptions about the input parameter uncertainties.

The simplest and most commonly used of these models is interval mathematics, which enables numerical computations in which each value is represented by a range of floating-point numbers without a probability structure. These intervals are processed by proper addition, subtraction, and/or multiplication operators, ensuring that each computed interval encloses the unknown value it represents.

Many analysts view interval mathematics as a subset of fuzzy theory, as interval variables can be viewed as a specific instance of fuzzy numbers. However, connecting interval mathematics to fuzzy set theory is not straightforward. Recently, fuzzy set theory and interval analysis have both been linked to a broader topological theory. Similarly, it is argued that fuzzy information granulation, rough set theory, and interval analysis are all subsets of a larger computational paradigm called granular computing, where these methodologies are complementary and symbiotic, rather than conflicting and exclusive.

This book focuses on the application of interval methods in uncertain power system analysis. In particular, after introducing the basic elements of interval computing, a set of conventional power system operation problems in the presence of data uncertainties are formalized and solved. Many numerical examples are presented and discussed in order to demonstrate the effectiveness of interval methods to reliably solve the complex problems encountered when analyzing the realistic operating scenarios of modern power systems.

1

Introduction to Reliable Computing

Many power-engineering computations, especially those related to system analysis, are affected by large and complex uncertainties. These uncertainties make it difficult to compute the "exact" problem solution, and the analyst is required to identify a proper approximated solution, which is as close as possible to the "exact" one. The difference between the "exact" and the approximated solution is commonly referred to as the solution "error."

The sources of uncertainties affecting power system analysis are multiple and heterogeneous, and can be both external and internal to the computing process. External uncertainties include measurement errors, missing data, and simplified models; while internal uncertainties are mainly related to the computing errors induced by digital processing, which frequently requires replacing rigorous mathematical models with discrete approximations (e.g. time discretization, truncation, and round-off errors).

Hence, when integrating the results of numerical analysis in power system operation tools, the impact of these uncertainties must be assessed and a formal error analysis should be considered an important part of the development process. The objective of such formal error analysis is to comprehensively assess the accuracy of all the computations involved, define the magnitude of the solution errors as a function of the values and the errors of the input data (i.e. variables and parameters).

Unfortunately, defining a formal mathematical process for specifying the accuracy of numerical computing algorithms is extremely complex, since estimating the propagation of both the external and internal errors for all the basic operations composing the computing process is often unfeasible, even for simple algorithms.

Moreover, accuracy specification requires the compliance of the input data with a set of strict prerequisites (e.g. well-conditioned matrices, no overflow,

Interval Methods for Uncertain Power System Analysis, First Edition. Alfredo Vaccaro.
© 2023 The Institute of Electrical and Electronics Engineers, Inc. Published 2023 by John Wiley & Sons, Inc.

and functions with bounded derivatives). Guaranteeing or even checking these prerequisites for all the possible combinations of the input data represents another challenging issue to address.

Hence, power system analysts frequently deploy numerical algorithms without formal accuracy specifications and rigorous error analysis, checking the consistency of the obtained results on the basis of their own experience or by crude tests. This practice could hinder the integration of approximate numerical computing in modern power systems tools, which are characterized by the presence of large data uncertainties, stemming from multiple and heterogeneous sources.

To try and overcome this limitation, the power system research community started adopting reliable computing-based models, which allow the accuracy of the computed quantities to be automatically estimated as part of the process of computing them. The application of these models in numerical computations is also referred to as self-validated computing, since it can estimate "a posteriori" the error magnitude of the entire computing process (Stolfi and De Figueiredo, 1997). This feature is extremely important, especially when the data uncertainties are induced by external causes. In this case, if the output errors computed by the self-validated model become too large, i.e. overcoming a fixed acceptable threshold, then specific remedial actions can be automatically triggered in order to enhance the model accuracy (e.g. acquire more data, re-adjourn the input parameters, and use more accurate models).

1.1 Elements of Reliable Computing

Let $f : R^m \rightarrow R^n$ be a continuous mathematical function, and suppose we need to compute $z = f(x)$ for $x \in R^m$. For this, we should implement a discrete numerical computation $Z = F(X)$, where X and Z are discrete mathematical objects approximating the corresponding continuous variables x and z.

To solve this issue different reliable computing models can be adopted, including probability distributions and range-based methods.

In particular, the adoption of probability distributions can approximate, in a statistical sense, the computed result Z by considering each component of the vector z as a real-value random variable, whose probability distribution function is frequently assumed to follow a Gaussian distribution. In this case, a reliable computing model should specify the statistical moments of each component z_i, and the corresponding covariance matrix describing the joint Gaussian probability distribution of the random vector $(z_1, \ldots, z_n)$, given those characterizing the random input variables $(x_1, \ldots, x_m)$.

The application of this probabilistic-based reliable computing model is often limited to specific application domains, which are characterized by Gaussian uncertainties, linear mappings, and negligible truncation errors. The lack of these conditions makes the statistical characterization of the computed result extremely complex or even mathematically intractable (Stolfi and De Figueiredo, 1997).

To try and overcome this limitation, most reliable computing models approximate the computed results by ranges, rather than by probability distributions.

According to these models, the approximated solution Z is described by means of its range $[Z]$, which is a compact set containing the "exact" solutions $z = f(x)$ for all the input variables x lying in the range $[X]$.

This important feature, which is usually referred to as the *fundamental invariant of range analysis*, guarantees that the range $[Z]$ contains the true solution set, provided that the input variables vary in a fixed range.

The simplest model of range analysis is Interval Arithmetic (Moore, 1966), which defines the range of each component of the computed result $[Z_i]$ by a real interval, which is a set of real numbers lying between its upper and lower bounds. Since no constraints relating these intervals are assumed, meaning that all the uncertainties are assumed to be statistically independent, the range of the computed result $[Z]$ is the Cartesian product of the ranges of its components $[Z_i]$, since all combinations of $z_1, \ldots, z_n$ in the box $[Z_1] \times \cdots \times [Z_n]$ are allowed.

In more advanced reliable computing-based models, such as those based on affine arithmetic (AA), the computed result also integrates useful information about the partial correlations between the output vector components z_i; hence, identifying the statistical dependencies between the input variables x_i and the computed result. In this case, the range of the computed result $[Z]$ is a proper subset of the Cartesian product of the individual ranges.

One of the most important features which characterizes all range-based models is their capability of computing, for every function $f : R^m \rightarrow R^n$, a *range extension* $F : \mathfrak{R}_m \rightarrow \mathfrak{R}_n$, which is characterized by the *fundamental invariant of range analysis*:

If the input vector $x = (x_1, \ldots, x_m)$ lies in the range jointly determined by the given approximate values $X = (X_1, \ldots, X_m)$, then the quantities $z = (z_1, \ldots, z_n) = f(x_1, \ldots, x_m)$ are guaranteed to lie in the range jointly defined by the approximate values $Z = (Z_1, \ldots, Z_n) = F(X_1, \ldots, X_m)$.

This property is extremely useful in reliable computing, since the joint range determined by the approximated values Z_i is an outer estimation of the real solution set, namely:

$$S = \{z : z = f(x_1, \ldots, x_m), x_1 \in X_1, \ldots, x_m \in X_m\} \subseteq Z \tag{1.1}$$

hence, introducing a conservative factor in approximating the solution vectors. This conservativism is a valuable feature of range analysis-based methods, which is extremely important in reliable power system analysis, since it allows the Analyst to bound all internal and external uncertainties in numerical computing. However, obtaining a suitable (not too large) conservativism level is a relevant problem in range-based computation, since the naive application of range-based models often results in extremely conservative solution ranges, which are too wide, and hence, not useful in realistic application domains.

Therefore, in order to assess and compare the conservativism of range-based reliable computing models, proper metrics should be defined. For this purpose, we introduce the *relative accuracy* of the range-based approximation Z, which is defined as:

$$r_a(Z) = \frac{\|Y\|}{\|[Z]\|} \tag{1.2}$$

where $\|Y\|$ and $\|[Z]\|$ are the norms of the real solution range $Y = \{f(x) : x \in [X]\}$ and the computed range $[Z]$, respectively.

This index is a reliable measure of the conservativism introduced by range-based computing, indicating the outer estimation accuracy by a number, which could vary between zero (i.e. the computed range is much wider than the real one) and one (i.e. no conservativism is introduced in range computing).

Obviously, this index requires that the input vector $x = (x_1, \dots, x_m)$ should vary in the range jointly determined by the given approximate values $X = (X_1, \dots, X_m)$, as stated by the fundamental invariant of range analysis. This hypothesis cannot be verified in real application domains, where some of the input data are acquired by measurements, which can be affected by unbounded errors. In this context, the assumption related to the inclusion of the output variables z in the range jointly determined by the approximate values Z should be revised by introducing probabilistic information about the range of the possible values $[Z]$. This requires determining, for the computed range $[Z]$, the corresponding *confidence range* p_Z, which is the probability of z to be included in $[Z]$. This approach differs from the Gaussian-based probabilistic models as it does not assume any hypothesis about the probability distribution of the error inside the range $[Z]$, but only that its integral over the computed range is at least p_Z.

Hence, the application of range-based methods in this context requires computing, for each mathematical operation, not only the joint range, but also a conservative estimation of the corresponding confidence level, as a function of the confidence level of the input data.

1.2   Interval Analysis

Interval analysis (IA), also referred to as interval arithmetic, is a range-based model for numerical computing that describes each uncertain variable x by an interval of real numbers $\bar{x}$, which is guaranteed to contain the (unknown) "true" value of x (Moore, 1966).

More specifically, a real interval is a compact set, which is defined as:

$$\bar{x} = [x_{lo}, x_{hi}] = \{x \in R : x_{lo} \leq x \leq x_{hi}\} \tag{1.3}$$

where x_{lo} and x_{hi} are the lower and upper bounds of the interval, respectively.

Those intervals are combined and processed by specific mathematical operators, which generalize all the real-value functions to process interval-based uncertain variables, in such a way that the computed intervals are guaranteed to contain all the possible values of the computed quantities. To achieve this, IA defines for each mathematical function $z = f(x_1, \dots, x_n)$ with $x_i \in R$ a proper *interval extension* $\bar{z} = \bar{f}(\bar{x}_1, \dots, \bar{x}_n)$, which allows computation of an interval $\bar{z}$ that contains all the values $z = f(x)$ for $(x_1, \dots, x_n)$ varying independently over the given intervals $(\bar{x}_1, \dots, \bar{x}_n)$.

Computing interval extension for linear functions is extremely simple, since it only requires defining a closed form expression for the extreme values of the function when its arguments vary independently over specified intervals.

On the contrary, computing interval extensions for nonlinear functions is a very complex issue. Indeed, defining analytic formula for computing the "exact" extreme values of these functions can be extremely difficult. Hence, in order to satisfy the fundamental invariant of range analysis, a conservative (but easily computable) approximation of the "exact" function range should be identified. In these cases, the corresponding computed intervals are outer estimations of the "exact" solution sets.

The interval extensions of the main mathematical operators and elementary functions can be composed in order to compute the interval extensions for any complex function by using the same mathematical schemes adopted in real numbers computing. Hence, any numerical algorithm processing real numbers can be automatically deployed for processing interval variables by replacing the real number operators and functions with their corresponding interval extensions. This scheme allows the numerical algorithms to evolve from computing with real numbers to computing with intervals and to identify solutions which are no longer deterministic (valid only for fixed real input data), but described by intervals that, according to the fundamental invariant of range analysis, are guaranteed to include the real values of the solutions for all instances of the uncertain input data (that should vary in the specified intervals).

1.3 Interval-Based Operators

The interval extensions of the basic mathematical operators can be easily defined as follows:

$$-\bar{x} = [-x_{hi}, -x_{lo}]$$

$$\bar{x} + \bar{y} = [x_{lo} + y_{lo}, x_{hi} + y_{hi}]$$

$$\bar{x} + c = [x_{lo} + c, x_{hi} + c]$$

$$\bar{x} - \bar{y} = [x_{lo} - y_{hi}, x_{hi} - y_{lo}] \tag{1.4}$$

$$c \cdot \bar{x} = [c \cdot x_{lo}, c \cdot x_{hi}] c > 0$$

$$c \cdot \bar{x} = [c \cdot x_{hi}, c \cdot x_{lo}] c < 0$$

To define the interval extension of the multiplication, $\bar{x} * \bar{y}$, it is necessary to derive an analytic expression of the minimum and maximum values of the function xy for (x, y) varying over the range $[x_{lo}, x_{hi}] \times [y_{lo}, y_{hi}]$. To solve this problem, it should be noted that the function xy is linear in y (x) for each fixed x (y); hence, its maximum and minimum values, referred to as a and b, respectively, are located at the corner of the rectangle $[x_{lo}, x_{hi}] \times [y_{lo}, y_{hi}]$, namely:

$$a = \min \{x_{lo} \cdot y_{lo}, x_{lo} \cdot y_{hi}, x_{hi} \cdot y_{lo}, x_{hi} \cdot y_{hi}\}$$

$$b = \max \{x_{lo} \cdot y_{lo}, x_{lo} \cdot y_{hi}, x_{hi} \cdot y_{lo}, x_{hi} \cdot y_{hi}\} \tag{1.5}$$

The same scheme can also be adopted for computing the interval extension of the division, $\bar{x}/\bar{y}$, which can be represented as the product of $\bar{x}$ by $1/\bar{y}$. In this case, it should be noted that the reciprocal function $1/y$ is not defined for $y = 0$; hence, this condition should be properly checked as follows:

$$\text{if } y_{lo} < 0 \text{ and } y_{hi} > 0 \text{ then } a = -\infty, b = +\infty$$

$$\text{if } y_{hi} = 0 \text{ then } a = -\infty$$

$$\text{if } y_{hi} \neq 0 \text{ then } a = \frac{1}{y_{hi}}$$

$$\text{if } y_{lo} = 0 \text{ then } b = +\infty \tag{1.6}$$

$$\text{if } y_{lo} \neq 0 \text{ then } b = \frac{1}{y_{lo}}$$

These interval-based arithmetic operators generalize the arithmetic of real numbers to the set of real intervals.

Other useful interval operators include the midpoint $m(\bar{x})$ and the radius $r(\bar{x})$ of the interval $\bar{x}$, which are defined as:

$$m(\bar{x}) = \frac{x_{lo} + x_{hi}}{2}$$
$$r(\bar{x}) = \frac{x_{hi} - x_{lo}}{2} \tag{1.7}$$

Moreover, it is also useful to define the intersection of two intervals $\bar{x}$ and $\bar{y}$, which is defined as:

$$\bar{x} \cap \bar{y} = [\max\{x_{lo}, y_{lo}\}, \min\{x_{hi}, y_{hi}\}] \tag{1.8}$$

and the convex hull of two intervals $\bar{x}$ and $\bar{y}$, which is the smallest interval $\bar{x} \cup \bar{y}$ containing $[\min\{x_{lo}, y_{lo}\}, \max\{x_{hi}, y_{hi}\}]$.

Finally, in order to solve multidimensional problems, we define interval matrices $\bar{\mathbf{A}} = (\bar{a}_{ij})$ with $i \in [1, m]$ and $j \in [1, n]$, and interval vectors $\bar{\mathbf{x}} = (\bar{x}_i)$, with $i \in [1, n]$, which are defined as follows (Götz and Günter, 2000):

$$[\mathbf{A}_{lo}, \mathbf{A}_{hi}] = \{\mathbf{B} \in R^{m \times n} : \mathbf{A}_{lo} \leq \mathbf{B} \leq \mathbf{A}_{hi}\}$$

$$[\mathbf{x}_{lo}, \mathbf{x}_{hi}] = \{\mathbf{z} \in R^n : \mathbf{x}_{lo} \leq \mathbf{z} \leq \mathbf{x}_{hi}\}$$

The real matrix $\mathbf{A} \in (R^{n \times n})$ is said to be an M matrix, if $a_{ij} \leq 0$ for $i \neq j$ and if $\mathbf{A}^{-1}$ exists and is nonnegative. This definition can be generalized to interval matrices; in particular, if each matrix $\mathbf{A}$ from a given interval matrix $\bar{\mathbf{A}}$ is an M matrix, then the interval matrix $\bar{\mathbf{A}}$ is an M matrix.

Starting from these definitions, it is possible to express the product $\bar{\mathbf{A}}\bar{\mathbf{x}}$ as follows:

$$\bar{\mathbf{A}}\bar{\mathbf{x}} = \sum_{j=1}^{n} \bar{a}_{ij}\bar{x}_j \subseteq \{\mathbf{A}\mathbf{x} : \mathbf{A} \in \bar{\mathbf{A}}, \mathbf{x} \in \bar{\mathbf{x}}\} \tag{1.9}$$

Hence, $\bar{\mathbf{A}}\bar{\mathbf{x}}$ is the interval vector containing the left set in (1.9).

An interval vector enclosing some set S as tight as possible is called the (interval) hull of S.

1.4 Interval Extensions of Elementary Functions

Computing the interval extension of the square root is straightforward since this function is monotonic, and a preliminary check of the input range is only required, as follows:

$$\text{if } x_{lo} < 0 \text{ and } x_{hi} < 0 \text{ then } \sqrt{\bar{x}} = [\ ]$$

$$\text{if } x_{lo} < 0 \text{ and } x_{hi} > 0 \text{ then } \sqrt{\bar{x}} = [0, \sqrt{x_{hi}}] \tag{1.10}$$

$$\text{if } x_{lo} > 0 \text{ and } x_{hi} > 0 \text{ then } \sqrt{\bar{x}} = [\sqrt{x_{lo}}, \sqrt{x_{hi}}]$$

A similar approach can be adopted for defining the interval extension of the logarithm log(x) (which is not defined at zero) and the exponential exp(x) = e^x, which are described in (1.11) and (1.12), respectively.

$$\text{if } x_{hi} < 0 \text{ then } \log \bar{x} = [\ ]$$

$$\text{if } x_{lo} < 0 \text{ and } x_{hi} > 0 \text{ then } \log \bar{x} = [-\infty, \log x_{hi}] \tag{1.11}$$

$$\text{if } x_{lo} > 0 \text{ and } x_{hi} > 0 \text{ then } \log \bar{x} = [\log x_{lo}, \log x_{hi}]$$

$$e^{\bar{x}} = [e^{x_{lo}}, e^{x_{hi}}] \tag{1.12}$$

The definition of interval extension of trigonometric functions is not trivial, because they could be nonmonotonic in the considered input interval, but this can be obtained by computing the maximum and minimum function values for inputs varying in the specified interval, which for cos and sin functions occur at integer multiples of π and $\pi/2$, respectively. Hence, the following procedures can be adopted for computing the corresponding interval extension of cos and sin functions, respectively:

$$\text{if } k\pi \in \bar{x} \ (k = 0, 2, 4, \dots) \text{ then } b = 1$$

$$\text{else } b = \max\left(\cos(x_{lo}), \cos(x_{hi})\right)$$

$$\text{if } k\pi \in \bar{x} \ (k = 1, 3, 5, \dots) \text{ then } a = -1 \tag{1.13}$$

$$\text{else } a = \min\left(\cos(x_{lo}), \cos(x_{hi})\right)$$

$$\cos \bar{x} = [a, b]$$

$$\text{if } k\frac{\pi}{2} \in \bar{x} \ (k = 0, 2, 4, \dots) \text{ then } b = 1$$

$$\text{else } b = \max\left(\sin(x_{lo}), \sin(x_{hi})\right)$$

$$\text{if } k\frac{\pi}{2} \in \bar{x} \ (k = 1, 3, 5, \dots) \text{ then } a = -1 \tag{1.14}$$

$$\text{else } a = \min\left(\sin(x_{lo}), \sin(x_{hi})\right)$$

$$\sin \bar{x} = [a, b]$$

These computing procedures can be properly generalized in order to compute the interval extension of any elementary function $f : D \subseteq R^n \to R^m$ with $\bar{x} \subseteq D$, which, thanks to the fundamental invariant of range analysis, is an enclosure of the corresponding function range.

The effectiveness of IA in conservatively estimating the range of any elementary function when the inputs ranges over assigned intervals is an effective tool, which can be used for effectively solving complex mathematical problems, such as:

1. Computing the global minima of scalar functions
2. Estimating the range of the Jacobian matrix

3. Verifying and enclosing solutions of initial value problems,
4. Computing the zeros of scalar functions.

1.5 Solving Systems of Linear Interval Equations

IA-based computing can be applied to solve the following uncertain linear problem:

$$\bar{A}\bar{x} = \bar{b} \tag{1.15}$$

where $\bar{A}$ and $\bar{b}$ are known $n \times n$ interval matrix, and n-dimension interval vector, respectively. These known quantities can be expressed as:

$$\bar{A} = [m(\mathbf{A}) - r(\mathbf{A}), m(\mathbf{A}) + r(\mathbf{A})]$$
$$\bar{b} = [m(\mathbf{b}) - r(\mathbf{b}), m(\mathbf{b}) + r(\mathbf{b})] \tag{1.16}$$

where $m(\mathbf{A})$ $(m(\mathbf{b}))$ and $r(\mathbf{A})$ $(r(\mathbf{b}))$ are the center and the radius of the interval matrix $\mathbf{A}$ (interval vector $\mathbf{b}$), respectively.

The solution set of these linear equations is defined as:

$$X = \{x \in R^n : Ax = b, A \in \mathbf{A}, b \in \mathbf{A}\} \tag{1.17}$$

Our task is to identify an outer interval enclosure of this set.
To this aim, let's define the following:

1. $e = (1, 1, \dots, l)^T \in R^n$
2. $f = -e,$
3. $Y = \{y : |y| = e, y \in R^n\}$, so that Y has 2^n elements. For example, for $n = 2$
   $Y = \{(1, 1); (-1, 1); (1, -1); (-1, -1)\}$
4. $\forall z \in R^n$, T_z is the diagonal matrix with the vector components z_i on its diagonal
   (namely $diag(T_z) = \{z_1, \dots, z_n\}$)
5. $A_{yz} = m(\mathbf{A}) - T_y r(\mathbf{A}) T_z$ and $b_y = m(\mathbf{b}) + T_y r(\mathbf{b})$ $\forall y, z \in R^n$
6. $d_y = m(\mathbf{A})^{-1} b_y$
7. $b_y = m(\mathbf{b}) + T_y r(\mathbf{b})$

Thanks to these definitions, it is possible to rigorously solve the problem (1.15) by solving the following 2^n linear complementary problems:

$$x^+ = A_{ye}^{-1} A_{yf} x^- + A_{ye}^{-1} b_y \quad \forall y \in Y \tag{1.18}$$

(2.4).

The solution of this problem for each $y \in Y$, here referred to as x_y, may be obtained using conventional techniques for solving linear complementarity problems. Once these 2^n solutions have been computed, the outer interval estimation of the solution set can be easily obtained as follows:

$$x_{lo} = \min(x_y : y \in Y)$$
$$x_{hi} = \max(x_y : y \in Y)$$

(1.19)

However, in order to simplify the solution process, i.e. by avoiding the inversion of the matrix A_{ye}, it is possible to recast the problem (1.18) as follows:

$$A_{yz}x = b_y$$
$$T_z x \geq 0$$
$$z \in Y$$

(1.20)

This problem can be solved by solving the systems $A_{yz}x = b_y$, for different z's until satisfying the condition $T_z x \geq 0$, which is equivalent to $z_j x_j \geq 0$ for each j. Hence, the following algorithm can be adopted to solve the problem (Rohn, 1989):

Algorithm

- *Step 0*: Select $z \in Y$ (recommended $z = sign(d_y)$)
- *Step 1*: Solve $A_{yz}x = b_y$
- *Step 2*: If $T_z x \geq 0$ set $x_Y = x$ and terminate
- *Step 3*: Otherwise, find

$$k = \min(j; z_j x_j < 0)$$

(1.21)

- *Step 4*: Set $z_k = -z_k$ and go to Step 1.

if A is regular, namely if each $A \in A$ is nonsingular, then this iterative scheme converges in a finite number of steps for each $y \in Y$ and for an arbitrary starting $z \in Y$ in step 0 (Rohn, 1989).

In many realistic cases, computing x_y only requires the solution of a single system $A_{yz}x = b_y$, and there is no need to compute all 2^n vectors yz; hence, a subset of Y can be considered in generating the vectors y.

Example 1.1 Let's consider the following set of linear interval equations:

$$[2,4]x_1 + [-2,-1]x_2 = [8,10]$$
$$[2,5]x_1 + [4,5]x_2 = [5,40]$$

(1.22)

In this case, it follows that:

$$m(\mathbf{A}) = \begin{bmatrix} 3 & -1.5 \\ 3.5 & 4.5 \end{bmatrix} \quad Y = \begin{bmatrix} 1 & 1 \\ -1 & 1 \\ 1 & -1 \\ -1 & -1 \end{bmatrix} \quad r(\mathbf{A}) = \begin{bmatrix} 1 & 0.5 \\ 1.5 & 0.5 \end{bmatrix}$$

(1.23)

$$m(\mathbf{b}) = [9, 22.5] \quad r(\mathbf{b}) = [1, 17.5]$$

and

$$x_y = \begin{bmatrix} 10 & 5 \\ 4 & 8 \\ 3.4615 & -3.0769 \\ 1.6154 & -0.7692 \end{bmatrix}$$

(1.24)

and the solution is

$$\bar{x}_1 = [1.6154, 10]$$
$$\bar{x}_2 = [-3.0769, 8]$$

(1.25)

Example 1.2 Let's consider the following set of linear interval equations:

$$[1, 1000]x_1 + [1, 1000]x_2 = [1, 2]$$
$$[-1000, -1]x_1 + [1, 1000]x_2 = [3, 4]$$

(1.26)

In this case, it follows that:

$$m(A) = \begin{bmatrix} 500.5 & 500.5 \\ -500.5 & 500.5 \end{bmatrix} Y = \begin{bmatrix} 1 & 1 \\ -1 & 1 \\ 1 & -1 \\ -1 & -1 \end{bmatrix} r(A) = \begin{bmatrix} 499.5000 & 499.5000 \\ 499.5000 & 499.5000 \end{bmatrix}$$

$$m(b) = [1.5, \ 3.5] \ r(b) = [0.5, \ 0.5]$$

(1.27)

and

$$x_y = \begin{bmatrix} -0.0020 & 3.9980 \\ -3.9950 & 0.0050 \\ 1.9950 & 0.0050 \\ -0.0020 & 0.0010 \end{bmatrix}$$

(1.28)

and the solution is

$$\bar{x}_1 = [-3.9950, 1.9950]$$
$$\bar{x}_2 = [0.0010, 3.9980]$$

(1.29)

Example 1.3

$$
m(\mathbf{A}) = \begin{bmatrix} 4.33 & -1.12 & -1.08 & 1.14 \\ -1.12 & 4.33 & 0.24 & -1.22 \\ -1.08 & 0.24 & 7.21 & -3.22 \\ 1.14 & -1.22 & -3.22 & 5.43 \end{bmatrix} \quad Y = \begin{bmatrix} 1 & 1 & 1 & 1 \\ 1 & 1 & 1 & -1 \\ 1 & 1 & -1 & 1 \\ 1 & 1 & -1 & -1 \\ 1 & -1 & 1 & 1 \\ 1 & -1 & 1 & -1 \\ 1 & -1 & -1 & 1 \\ 1 & -1 & -1 & -1 \\ -1 & 1 & 1 & 1 \\ -1 & 1 & 1 & -1 \\ -1 & 1 & -1 & 1 \\ -1 & 1 & -1 & -1 \\ -1 & -1 & 1 & 1 \\ -1 & -1 & 1 & -1 \\ -1 & -1 & -1 & 1 \\ -1 & -1 & -1 & -1 \end{bmatrix} \tag{1.30}
$$

$$
r(\mathbf{A}) = \begin{bmatrix} 0.005 & 0.005 & 0.005 & 0.005 \\ 0.005 & 0.005 & 0.005 & 0.005 \\ 0.005 & 0.005 & 0.005 & 0.005 \\ 0.005 & 0.005 & 0.005 & 0.005 \end{bmatrix}
$$

$m(\mathbf{b}) = [3.52, \ 1.57, \ 0.54, -1.09] \quad r(\mathbf{b}) = [0.005, \ 0.005, \ 0.005, \ 0.005]$

$$
x_y = \begin{bmatrix} 1.0509 & 0.5689 & 0.1164 & -0.2218 \\ 1.0517 & 0.5670 & 0.1129 & -0.2299 \\ 1.0502 & 0.5680 & 0.1106 & -0.2253 \\ 1.0510 & 0.5662 & 0.1072 & -0.2333 \\ 1.0492 & 0.5611 & 0.1155 & -0.2237 \\ 1.0500 & 0.5592 & 0.1121 & -0.2318 \\ 1.0485 & 0.5603 & 0.1098 & -0.2271 \\ 1.0492 & 0.5584 & 0.1064 & -0.2352 \\ 1.0433 & 0.5671 & 0.1156 & -0.2211 \\ 1.0440 & 0.5653 & 0.1122 & -0.2291 \\ 1.0425 & 0.5663 & 0.1099 & -0.2245 \\ 1.0433 & 0.5644 & 0.1065 & -0.2325 \\ 1.0416 & 0.5594 & 0.1148 & -0.2230 \\ 1.0423 & 0.5576 & 0.1114 & -0.2310 \\ 1.0408 & 0.5586 & 0.1091 & -0.2264 \\ 1.0416 & 0.5567 & 0.1057 & -0.2344 \end{bmatrix} \tag{1.31}
$$

and the solution is

$$\bar{x}_1 = [1.0408, 1.0517]$$

$$\bar{x}_2 = [0.5567, 0.5689]$$

$$\bar{x}_3 = [0.1057, 0.1164]$$

$$\bar{x}_4 = [-0.2352, -0.2211]$$

(1.32)

1.6 Finding Zeros of Nonlinear Equations

IA-based operators can be deployed to derive simple and effective tests aimed at checking if a continuous and differentiable nonlinear function $f : D \subseteq R^n \rightarrow R^n$ has at least a zero x^* in the interval $\bar{x} \subseteq D$, and computing a tighter interval enclosure $\bar{x}^*$ containing x^* (e.g. $r(\bar{x}^*) \leq r(\bar{x})$).

Different IA-based methods can be adopted for solving these problems by eliminating from the initial interval $\bar{x} \subseteq D$ those regions that do not contain any solution. Hence, the initial interval is divided and contracted in such a way that the subintervals containing all the solutions are identified.

The simplest example of the IA-based contraction technique for computing a reliable enclosure of all the zeros of a nonlinear function in a fixed interval is based on the following iterative scheme (Götz and Günter, 2000), which is valid for any continuous function $f : D \subseteq R^n \rightarrow R^n$ such that $f([x]) \subseteq \bar{x} \subseteq D$:

$$\bar{x}^0 = \bar{x}$$

$$\bar{x}^{k+1} = f(\bar{x}^k) \; k = 0, 1, \ldots$$

(1.33)

which iteratively contracts the solution enclosure $\bar{x}^{k+1} \subseteq \bar{x}^k \cdots \subseteq \bar{x}^0 = \bar{x}$, hence, converging to the interval $\bar{x}^*$ containing all the zeros of the function f in $\bar{x}$.

Example 1.4 Consider the problem of finding the enclosure of the zeros of the function $\sin(x)^2 + 2 - x = 0$ in the interval $\bar{x} = [-4, 4]$ (the exact solution is $x^* = 2.4282$)

$$\bar{x}^1 = [2.0000, 3.0000]$$
$$\bar{x}^2 = [2.0199, 2.8269]$$
$$\bar{x}^3 = [2.0958, 2.8115]$$
$$\bar{x}^4 = [2.1050, 2.7488]$$
$$\bar{x}^5 = [2.1465, 2.7408]$$
$$\bar{x}^6 = [2.1522, 2.7036]$$
$$\bar{x}^7 = [2.1799, 2.6984]$$
$$\bar{x}^8 = [2.1839, 2.6727]$$
$$\bar{x}^9 = [2.2042, 2.6689]$$
$$\bar{x}^{10} = [2.2073, 2.6497]$$

(1.34)

An alternative approach for finding the interval enclosure of all zeros of a generic set on continuous nonlinear equations is based on the following theorem:

Theorem 1.1   *Let $f : D \subseteq R^n \rightarrow R^n$ continuous, $\bar{x} \subseteq D$, and*

$$\bar{x}_{lo}^i = (\bar{x}_1, \dots, \bar{x}_{i-1}, x_{i_{lo}}, \bar{x}_{i+1}, \dots, \bar{x}_n)^T$$

$$\bar{x}_{hi}^i = (\bar{x}_1, \dots, \bar{x}_{i-1}, x_{i_{hi}}, \bar{x}_{i+1}, \dots, \bar{x}_n)^T$$

then, if

$$f_{i_{hi}}(\bar{x}_{lo}^i) \le 0, f_{i_{lo}}(\bar{x}_{hi}^i) \ge 0,$$

or

$$f_{i_{lo}}(\bar{x}_{lo}^i) \ge 0, f_{i_{hi}}(\bar{x}_{hi}^i) \le 0$$

holds for each $i = 1, \dots, n$, then f has at least one zero in $\bar{x}$.

By using this simple and effective test, it is possible to subdivide and contract the initial interval $\bar{x}$, selecting only those subintervals containing the problem solutions.

1.7   Solution of Systems of Nonlinear Interval Equations

Another useful application of IA-based computing is in computing reliable interval enclosures of the solution sets of uncertain nonlinear equations, which can be formalized as follows:

$$\mathbf{f}(\mathbf{x}, p_1, \dots, p_m) = 0 \tag{1.35}$$

where $(p_1, \dots, p_m)$ denotes a set of uncertain parameters, whose values are unknown but bounded in the following intervals: $\bar{\mathbf{p}} = (\bar{p}_1, \dots, \bar{p}_m)$.

Hence, the range of $\mathbf{f}$ for each $\mathbf{x} \in R^n$ can be estimated by computing the interval-valued function $\mathbf{F}(\mathbf{x}, \bar{\mathbf{p}})$, which represents the interval-extension of $\mathbf{f}$, and the overall problem can be solved by identifying a proper interval enclosure of the solution set of the uncertain system of uncertain equations $\mathbf{F}(\mathbf{x}, \bar{\mathbf{p}}) = 0$, which is defined as follows:

$$S = \{\mathbf{x} : \mathbf{f}(\mathbf{x}, \mathbf{p}) = 0, \mathbf{p} \in \bar{\mathbf{p}}\} \tag{1.36}$$

To enclose this solution set, it is possible to adopt iterative contraction schemes based on Krawczyk and interval-Newton operators, which allow testing for the existence of problem solutions in given intervals (Gwaltney et al., 2008), and contracting the solution bounds by discarding those intervals that do not include any solutions. According to this iterative branch and bound technique, the problem solutions can be obtained by splitting the given intervals (e.g. by using a bisection scheme), and sequentially testing for the existence of problem solutions in the resulting subintervals Gwaltney et al. (2008).

In particular, let $\bar{x}(k)$ denote the interval to be tested at the kth iteration step, by using the Krawczyk method, it is possible to contract this interval by using the following equation (Götz and Günter, 2000):

$$\bar{x}(k + 1) = \bar{x}(k) \cap \bar{K}(k)$$

$$\bar{K}(k) = x(k) - Y(k)f(x(k)) + (I - Y(k)\bar{J}(\bar{x}(k)))(\bar{x}(k) - x(k)) \tag{1.37}$$

where $\bar{J}(\bar{x}(k))$ is the interval extension of the Jacobian matrix of $f(x)$, $x(k) \in \bar{x}(k)$ is an internal point of $\bar{x}(k)$ (e.g. its mid-point), and $Y(k)$ is a real preconditioning matrix, which is frequently chosen as the inverse of the midpoint of $\bar{J}(\bar{x}(k))$, or the inverse of $J(m(\bar{x}(k)))$. One of the most interesting features of the Krawczyk method is that any problem solutions included in $\bar{x}(k)$ are also included in $\bar{K}(k)$, hence $\bar{x}(k) \subset \bar{x}(k + 1)$, and if $\bar{x}(k) \cap \bar{K}(k) = \emptyset$, then $\bar{x}(k)$ does not include any solutions and can be discharged (Gwaltney et al., 2008). Moreover, if $\bar{x}(k)$ cannot be discharged or contracted, then it should be subdivided in two sub-intervals, repeating the procedure for both of them.

The same contraction scheme can be implemented by using the interval-Newton method. In this case, the solution iteration requires the solution of a set of linear interval equations:

$$\bar{x}(k + 1) = \bar{x}(k) \cap \bar{N}(k)$$

$$Y(k)\bar{J}(\bar{x}(k))(\bar{N}(k) - x(k)) = -Y(k)f(x(k)) \tag{1.38}$$

Also in this case, any solutions included in $\bar{x}(k)$ are also included in $\bar{N}(k)$; hence, $\bar{x}(k) \subset \bar{x}(k + 1)$, and if $\bar{x}(k) \cap \bar{N}(k) = \emptyset$, then $\bar{x}(k)$ does not include any solutions and can be eliminated Gwaltney et al., (2008). Moreover, if $\bar{x}(k)$ cannot be contracted, then it is bisected, and the procedure applied for each resulting subinterval.

The application of these contraction techniques allows the computation of interval enclosures of the solution set S. But it is important to note that since S is often not an interval, these interval enclosures tend to overestimate the real solution set. To reduce this overestimation, it is possible to split the set of uncertain parameters $\bar{p}$ into subintervals, computing an interval enclosure of the solution set for each obtained subinterval, and computing the final solution as the union of these subsolution sets.

An alternative approach for solving the problem formulated in (1.35) is based on the application of the multidimensional interval Newton method (Götz and

Günter, 2000). In particular, let $IGA(\bar{\mathbf{A}}, \bar{\mathbf{b}})$ be the solution of the following system of linear interval equations:

$$\bar{\mathbf{A}}\bar{\mathbf{x}} = \bar{\mathbf{b}} \tag{1.39}$$

which can be obtained by applying one of the techniques described in Section 1.5. Hence, it follows that:

$$\{x = A^{-1}b : \mathbf{A} \in \bar{\mathbf{A}}; \mathbf{b} \in \bar{\mathbf{b}}\} \subset IGA(\bar{\mathbf{A}}, \bar{\mathbf{b}}) \tag{1.40}$$

where $IGA(\bar{\mathbf{A}})$ denotes an interval enclosure for the inverses of all matrices $A \in \bar{\mathbf{A}}$, or, equivalently, the interval matrix whose ith column is obtained as $IGA(\bar{\mathbf{A}}; e^i)$ and e^i is the ith unit vector.

If $\mathbf{f}$ is continuously differentiable, then:

$$\mathbf{f}(\mathbf{x}) - \mathbf{f}(\mathbf{y}) = \mathbf{g}(\mathbf{y}, \mathbf{x})(\mathbf{x} - \mathbf{y})$$

$$\mathbf{g}(\mathbf{y}, \mathbf{x}) = \int_0^1 \mathbf{J}(\mathbf{y} + t(\mathbf{x} - \mathbf{y}))dt \tag{1.41}$$

$$\forall x, y \in \bar{\mathbf{x}}$$

Consequently, it results that:

$$\mathbf{g}(\mathbf{y}, \mathbf{x}) = \mathbf{g}(\mathbf{x}, \mathbf{y})$$

$$\mathbf{y} + t(\mathbf{x} - \mathbf{y}) \in \bar{\mathbf{x}} \quad \forall t \in [0, 1] \tag{1.42}$$

$$\mathbf{g}(\mathbf{y}, \mathbf{x}) \in \bar{\mathbf{J}}(\bar{\mathbf{x}})$$

where $\bar{\mathbf{J}}(\bar{\mathbf{x}})$ is the interval extension of the Jacobian of $\mathbf{f}$. Hence, for each $\mathbf{y} \in \bar{\mathbf{x}}$, it follows that:

$$\mathbf{x} - \mathbf{J}^{-1}(\mathbf{y}, \mathbf{x})\mathbf{f}(\mathbf{x}) = \mathbf{y} - \mathbf{J}^{-1}(\mathbf{y}, \mathbf{x})\mathbf{f}(\mathbf{y}) \in \mathbf{y} - IGA(\bar{\mathbf{J}}(\bar{\mathbf{x}}), \mathbf{f}(\mathbf{y})) \tag{1.43}$$

thanks to this equation, if $\mathbf{x} \in \bar{\mathbf{x}}$ is a zero of $\mathbf{f}$, then:

$$\mathbf{x} \in \mathbf{y} - IGA(\bar{\mathbf{J}}(\bar{\mathbf{x}}), \mathbf{f}(\mathbf{y})) \tag{1.44}$$

Starting from this result, the following interval Newton-based solution scheme can be applied for solving the problem formulated in (1.15):

$$\bar{\mathbf{x}}(k + 1) = \bar{\mathbf{x}}(k) \cap \bar{\mathbf{N}}_1(k)$$

$$\bar{\mathbf{N}}_1(k) = m(\bar{\mathbf{x}}(k)) - IGA(\bar{\mathbf{J}}(\bar{\mathbf{x}}(k)), f(m(\bar{\mathbf{x}}(k)))) \tag{1.45}$$

Example 1.5 Let's solve the following set on nonlinear interval equations:

$$x_1^2 + x_2 = [10.9, 11.1] \tag{1.46}$$

$$x_1 + x_2^2 = [6.9, 7.1] \tag{1.47}$$

The rigorous solution of this problem is:

$$\bar{x}_1 = [2.9782, 3.0217] \tag{1.48}$$

$$\bar{x}_2 = [1.9693, 2.0302] \tag{1.49}$$

The application of the previously described solution scheme results in the following iterations:

1. $\bar{x}_1(0) = [2.5, 3.5]$ $\bar{x}_2(0) = [1.5, 2.5]$
2. $\bar{x}_1(1) = [2.8695, 3.1305]$ $\bar{x}_2(1) = [1.8173, 2.1827]$
3. $\bar{x}_1(2) = [2.9694, 3.0306]$ $\bar{x}_2(2) = [1.9506, 2.0494]$
4. $\bar{x}_1(3) = [2.9777, 3.0223]$ $\bar{x}_2(3) = [1.9682, 2.0318]$
5. $\bar{x}_1(4) = [2.9780, 3.0220]$ $\bar{x}_2(4) = [1.9689, 2.0311]$
6. $\bar{x}_1(5) = [2.9780, 3.0220]$ $\bar{x}_2(5) = [1.9690, 2.0310]$

Example 1.6 Let's solve the following set on nonlinear interval equations:

$$x_1^2 + x_2 = [36.85, 37.05] \tag{1.50}$$

$$x_1 + x_2^2 = [6.95, 7.1] \tag{1.51}$$

$$x_1 + x_2 + x_3 = [10.9, 11.15] \tag{1.52}$$

The rigorous solution of this problem is:

$$\bar{x}_1 = [5.9827, 6.0066] \tag{1.53}$$

$$\bar{x}_2 = [0.9713, 1.0570] \tag{1.54}$$

$$\bar{x}_3 = [3.8511, 4.1806] \tag{1.55}$$

The application of the previously described solution scheme results in the following iterations:

1. $\bar{x}_1(0) = [5, 6.5]$ $\bar{x}_2(0) = [0.5, 1.5]$ $\bar{x}_3(0) = [3, 4.5]$
2. $\bar{x}_1(1) = [5.8624, 6.1376]$ $\bar{x}_2(1) = [0.6562, 1.3688]$ $\bar{x}_3(1) = [3.5573, 4.4677]$
3. $\bar{x}_1(2) = [5.9684, 6.0207]$ $\bar{x}_2(2) = [0.8398, 1.1904]$ $\bar{x}_3(2) = [3.7290, 4.3017]$
4. $\bar{x}_1(3) = [5.9799, 6.0093]$ $\bar{x}_2(3) = [0.9406, 1.0896]$ $\bar{x}_3(3) = [3.8216, 4.2091]$
5. $\bar{x}_1(4) = [5.9821, 6.0070]$ $\bar{x}_2(4) = [0.9665, 1.0637]$ $\bar{x}_3(4) = [3.8453, 4.1853]$
6. $\bar{x}_1(5) = [5.9824, 6.0068]$ $\bar{x}_2(5) = [0.9698, 1.0604]$ $\bar{x}_3(5) = [3.8483, 4.1823]$
7. $\bar{x}_1(6) = [5.9824, 6.0067]$ $\bar{x}_2(6) = [0.9701, 1.0601]$ $\bar{x}_3(6) = [3.8486, 4.1820]$
8. $\bar{x}_1(7) = [5.9824, 6.0067]$ $\bar{x}_2(7) = [0.9702, 1.0600]$ $\bar{x}_3(7) = [3.8486, 4.1820]$

1.8 The Overestimation Problem

The main problem deriving from the application of interval arithmetic-based computing in solving uncertain mathematical programming problems is the overestimation of the real solution sets, which could be extremely large and, in some cases, useless. This limitation is particularly severe in iterative schemes, where the solution computed at one iteration is processed as the input of the next iteration, which is frequently adopted in uncertain power system analysis.

This limitation mainly derives from the fundamental assumption of IA, which assumes all the uncertain variables to vary *independently* over fixed intervals. Hence, if this assumption is not valid, meaning that there are statistical correlations between the values assumed by the uncertain variables, then some combinations of these values in the given intervals are not valid, and the solution bounds computed by using interval-based operators could be wider than the exact ones.

An extreme consequence of this phenomenon, which is known as the *dependency problem*, can be observed by evaluating the expression $x - x$ by interval arithmetic-based operators. For example, by computing this expression for the interval $\bar{x} = [2, 5]$, we get $[2 - 3, 5 - 2] = [-3, 3]$ instead of $[0, 0]$, which is the true range of the expression. This overestimation derives from the fact that the interval arithmetic-based operators assume that each interval represents an independent uncertain variable, while in the considered example, the two intervals refer to the same uncertain variable x.

This dependency problem generally affects the conservativism of all the interval-based operators processing two or more uncertain variables that are not independent. In this case, the uncertain variables are erroneously correlated, and the computed intervals may be much wider than the exact bounds.

In particular, let's consider the expression $x(10 - x)$, where x is an uncertain variable ranging in the interval $\bar{x} = [4, 6]$. The application of the interval operators defined in Section 1.3 computes the range of this expression by using the following procedure:

$$10 - \bar{x} = [4, 6]$$

$$\bar{x}(10 - \bar{x}) = [4, 6] \cdot [4, 6] = [16, 36]$$

(1.56)

Then, considering that the exact range of $x(10 - x)$ is $[24, 25]$, it can be concluded that the relative accuracy of the computed range is $(25 - 24)/(36 - 16) = 0.05$, hence, it is 20 times wider than the real one. This large overestimation is a result of the intrinsic correlation between the uncertain quantities x and $10 - x$, which is not modeled by the interval-based multiplication operator.

The consequences of this over-estimation phenomenon are particularly severe in iterative computing schemes, where the relative accuracy of the computed quantity at each stage tends to the product of the relative accuracy of all the previous iterations, which could diverge at an exponential rate generating the so-called "error explosion" phenomenon. In this case, after a few iterations, the computed intervals may become extremely wide and, consequently, useless.

Consider, for example, the following first-order ordinary differential equation:

$$\frac{dy}{dt} + \bar{p}y(t) = c \qquad (1.57)$$

where $\bar{p} = [1, 1.1]$ and $c = 1$. By applying a conventional Euler-based solution scheme and the previously described IA-based operators, the trajectories reported in Figure 1.1 are obtained.

To mitigate the over-estimation problem of interval arithmetic in computing the range of functions, namely $\bar{f}(\bar{x}, \bar{y}, \dots)$, it is possible to split the intervals $\bar{x} \times \bar{y} \times \cdots$ into subranges, computing the function range on each of these and properly combining the computed results. However, this approach could become ineffective when the relative accuracy of the adopted mathematical operators is independent of the width of the input interval variables, which frequently happens when solving realistic problems. In particular, if the relative accuracy of an interval-based computation is unacceptable by a factor of 100, then, in order to compute a useful result, the domain of the input interval variables must be split into 100 subintervals.

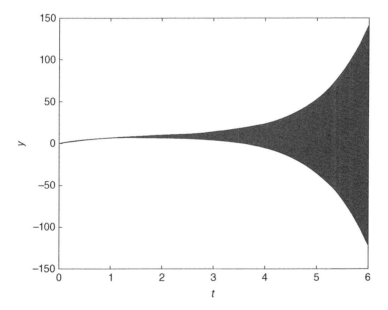

Figure 1.1 The error explosion problem.

A more effective technique for mitigating the error explosion problem is the re-formulation of the problem equations in order to reduce unfavorable correlations between the arguments of interval-based operations, e.g. by reducing the number of occurrences of each interval variable.

Another effective method frequently adopted in IA-based computing is to collect multiple interval-based operations into a single operation, defining a procedure for reliably computing the range of this macrooperation by explicitly considering the correlation between the input variables of all the elementary operations. Frequently, this approach allows computing tighter bounds compared to the sequential application of the elementary interval-based operations, typically referred to as "naive" interval arithmetic, but it can be applied only to simple uncertain computations. Simple uncertain computations are characterized by a limited number of interval variables, which are, in turn, characterized by restricted uncertain domains.

An interesting example showing the effects of the error explosion problem is the evaluation of powers $z \leftarrow x^n$ by applying the conventional interval-based multiplication operator. As confirmed in the following example, this computation returns over-conservative results, especially in the presence of interval variables close to zero.

Example 1.7 $z = f(x) = x^2$, for $\bar{x} = [-2, 2], \bar{z} = [-4, 4]$, even though x^2 cannot be negative.

Another typical overestimation problem derives from the evaluation of polynomials over an interval by sequentially deploying the sum and multiplication interval-based operators. For example the "naive" evaluation of the polynomial $f(x) = \sum_{i=0}^{\infty} a_i x^i$, when some a_i are negative, exhibits poor accuracy, even if the series is strongly convergent. This is mainly due to the cumulative effects of the negative correlations between the input variables and the computed results.

To overcome the overestimation of "naive" interval arithmetic, mitigating the error explosion problem, more sophisticated reliable computing techniques aimed at modeling the correlations between the uncertain variables should be deployed.

1.9 Affine Arithmetic

AA is a reliable computing model, which solves uncertain mathematical problems by computing outer estimations of the solution domains, modeling the correlations between the uncertain variables. This enhanced modeling feature makes the solution bounds computed by AA-based processing much tighter than those computed by "naive" IA; hence, resulting in approximation errors that are approximately quadratic in the uncertainty of the input variables. This feature is extremely useful in avoiding the error explosion problem.

In AA each uncertain variable x is described by a first-degree polynomial, which is called the *affine form* $\hat{x}$:

$$\hat{x} = x_0 + x_1\varepsilon_1 + x_2\varepsilon_2 + \cdots + x_n\varepsilon_n \tag{1.58}$$

where the coefficients x_0 and $x_1, \ldots, x_n$ are known real numbers, which are denoted as the *central value* and the *partial deviations* of $\hat{x}$, respectively, while ε_i are random variables uniformly distributed in the interval $[-1, 1]$, which are denoted as the *noise symbols* of $\hat{x}$. In particular, each noise symbol ε_i describes an independent source of uncertainty affecting the uncertain variable x, which could be either exogenous (e.g. measurement/estimation errors, forecasting errors) or endogenous (round-off errors, numerical approximation), while the partial deviation x_i describes the corresponding magnitude.

One of the most important features which characterizes AA is the *fundamental invariant of affine arithmetic*, which states that for any value assumed by the random variable x, there is a unique instance of the noise variables which makes the affine form $\hat{x}$ equal to this value:

$$\forall x_f \; \exists! \; (\varepsilon_{1f}, \ldots, \varepsilon_{nf}) \in ([-1, 1], \ldots, [-1, 1]) :$$
$$x_f = x_0 + x_1\varepsilon_{1f} + x_2\varepsilon_{2f} + \cdots + x_n\varepsilon_{nf} \tag{1.59}$$

The importance of this feature is much more relevant considering that the noise symbols can be shared by several affine forms, which allow modeling the impact of the same uncertainty source on different random variables. Indeed, if two random variables x and y are affected by the same uncertainty source, then the corresponding partial dependency characterizing these variables can be modeled in AA by defining two affine forms $\hat{x}$ and $\hat{y}$, which share the same noise symbol ε_k, determining proper values of the corresponding partial deviations x_k and y_k in order to define the sign and magnitude of the dependency.

For example, assume that two random variables x and y are represented by the following affine forms:

$$x = 5 + \varepsilon_1 + 2\varepsilon_2 - \varepsilon_4$$
$$y = 10 - 2\varepsilon_1 + 3\varepsilon_3 + 4\varepsilon_4 \tag{1.60}$$

By analyzing these affine forms, it is possible to infer that $x \in \bar{x} = [1, 9]$ and $y \in \bar{y} = [1, 19]$. Moreover, since these affine forms share some noise symbols, namely ε_1 and ε_4, it can be argued that the variables x and y are correlated, since they are affected by two common sources of uncertainty. Hence, all the possible instances of these variables are constrained in the dark gray region indicated in Figure 1.2, which significantly differs from the rectangle $[1, 9] \times [1, 19]$.

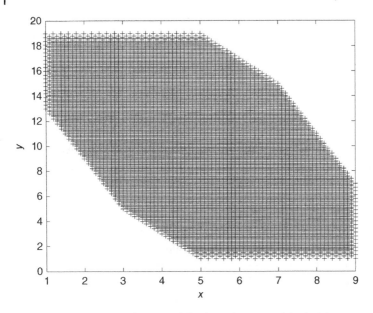

Figure 1.2 Joint range of two partially dependent quantities in AA.

This important and useful information represents the main advantage of using AA compared to IA. Indeed, the interval variables $\bar{x}$ and $\bar{y}$ do not allow any information about the partial dependency between the variables x and y to be inferred, since the joint range of $\bar{x}$ and $\bar{y}$ is $[1, 9] \times [1, 19]$, which is a conservative (outer) estimation of the real domain. Differently from IA, AA approximates the solution domain by a convex polygon, which is symmetric around (x_0, y_0). The shape and size of this polygon depends on the number of noise symbols and the value of the partial deviations, respectively. In particular, each noise symbol ε_i characterizing $\hat{x}$ or $\hat{y}$ corresponds to a pair of parallel sides, whose direction depends by the corresponding partial deviations, which defines the vectors $(2x_i, 2y_i)$ and $(-2x_i, -2y_i)$. In particular, in the previous example, the joint range is represented by a polygon composed of four pairs of parallel sides, each one corresponding to one of the four noise symbols, whose directions are identified by the following vectors:

$$V_1 = (2x_1, 2y_1), (-2x_1, -2y_1)$$
$$V_2 = (2x_1, 2y_1), (-2x_1, -2y_1)$$
$$V_3 = (2x_1, 2y_1), (-2x_1, -2y_1)$$
$$V_4 = (2x_1, 2y_1), (-2x_1, -2y_1)$$

(1.61)

Each point (x, y) lying on the ith parallel side corresponds to the value of $\hat{x}$ or $\hat{y}$ obtained by varying ε_i over $[-1, 1]$, and by fixing all the other noise symbols ε_j with $j \neq i$ to ± 1.

The geometric interpretation of the joint range of two affine forms can be generalized to the case of m affine forms with n noise symbols. In this case, the joint range of the affine forms is described by a center-symmetric convex polytope in R^m, which is the parallel projection of the hyperbox U^n by the multidimensional mapping described by the m affine forms.

1.9.1 Conversion Between AA and IA

An affine form $\hat{x}$ can be converted to an interval variable $\bar{x}$ according to the following equation:

$$\bar{x} = [x_0 - r(\hat{x}), x_0 + r(\hat{x})] \tag{1.62}$$

where $r(\hat{x}) = x_0 + \sum_{i=1}^{n} |x_i|$ is known as the radius or the total deviation of $\hat{x}$, and $\bar{x}$ is the smallest interval containing all the values assumed by $\hat{x}$ when each noise symbol ε_i varies independently over $[-1, 1]$.

As expected, the interval conversion of affine forms does not model any dependency between the uncertain variables, which are modeled as independent random variables uniformly distributed in the computed intervals.

The conversion of an interval variable $\bar{x} = [a, b]$ into an equivalent affine form can be obtained as the following:

$$\hat{x} = x_0 + x_k \varepsilon_k \tag{1.63}$$

where:

$$x_0 = \frac{a + b}{2}$$
$$x_k = \frac{b - a}{2} \tag{1.64}$$

Observe that the noise symbol ε_k models the overall uncertainty affecting the random variable x and, since the interval variable $\bar{x}$ does not describe any dependency information with other variables, it should be different from all the other defined noise symbols.

1.9.2 AA-Based Operators

In order to solve uncertain mathematical problems by AA, it is necessary to define proper mathematical operators, which extend all the basic operations on real numbers to affine forms.

In particular, given two affine forms:

$$\hat{x} = x_0 + x_1 \varepsilon_1 + x_2 \varepsilon_2 + \cdots + x_n \varepsilon_n$$
$$\hat{y} = y_0 + y_1 \varepsilon_1 + y_2 \varepsilon_2 + \cdots + y_n \varepsilon_n \tag{1.65}$$

and a generic binary operator on real numbers $f(x,y) \to z$, the corresponding AA counterpart, which is indicated as $\hat{f}(\hat{x}, \hat{y}) \to \hat{z}$, is a procedure that allows computing the affine form $\hat{z}$ such that $\forall (\tilde{\varepsilon}_1, \tilde{\varepsilon}_2, \ldots, \tilde{\varepsilon}_n) \in U^n$:

$$z = f(x, y)$$
$$= f(x_0 + x_1\tilde{\varepsilon}_1 + x_2\tilde{\varepsilon}_2 + \cdots + x_n\tilde{\varepsilon}_n, y_0 + y_1\tilde{\varepsilon}_1 + y_2\tilde{\varepsilon}_2 + \cdots + y_n\tilde{\varepsilon}_n)$$
$$\in [z_0 - r(\hat{z}), z_0 + r(\hat{z})] \tag{1.66}$$

Hence, the overall problem is to compute the parameters of the affine form $\hat{z}$, namely the central value z_0 and the partial deviations z_i, by preserving all the correlations between the variables x, y, and z.

In particular, if the function f is linear, then $\hat{z}$ can be expressed by combining the same noise symbols characterizing $\hat{x}$ and $\hat{y}$, which will be referred to as the *primitive noise symbols*, namely:

$$\hat{z} = z_0 + z_1\varepsilon_1 + z_2\varepsilon_2 + \cdots + z_n\varepsilon_n \tag{1.67}$$

and the corresponding parameters z_0 and z_i can be easily computed by directly manipulating (1.66). In particular, for any $\alpha, \zeta \in R$:

$$\hat{x} \pm \hat{y} = (x_0 \pm y_0) + (x_1 \pm y_1)\varepsilon_1 + (x_2 \pm y_2)\varepsilon_2 + \cdots + (x_n \pm y_n)\varepsilon_n$$
$$\alpha\hat{x} = (\alpha x_0) + (\alpha x_1)\varepsilon_1 + (\alpha x_2)\varepsilon_2 + \cdots + (\alpha x_n)\varepsilon_n \tag{1.68}$$
$$\zeta \pm \hat{x} = (\zeta \pm x_0) + x_1\varepsilon_1 + x_2\varepsilon_2 + \cdots + x_n\varepsilon_n$$

In these cases, $\hat{z}$ exactly describes all the correlations characterizing the uncertain variables x, y, and z, and is coherent with the corresponding affine forms $\hat{x}, \hat{y}$, namely:

$$\forall (\tilde{\varepsilon}_1, \tilde{\varepsilon}_2, \ldots, \tilde{\varepsilon}_n) \in U^n :$$
$$z_0 + z_1\tilde{\varepsilon}_1 + z_2\tilde{\varepsilon}_2 + \cdots + z_n\tilde{\varepsilon}_n$$
$$= f(x_0 + x_1\tilde{\varepsilon}_1 + x_2\tilde{\varepsilon}_2 + \cdots + x_n\tilde{\varepsilon}_n, y_0 + y_1\tilde{\varepsilon}_1 + y_2\tilde{\varepsilon}_2 + \cdots + y_n\tilde{\varepsilon}_n)$$
$$\tag{1.69}$$

Thanks to this important feature, it is possible to argue that $\hat{x} - \hat{x}$ is rigorously zero, since the AA-based subtraction formula recognizes that the input operators are characterized by the same central values and partial deviations, hence describing the same variable. Moreover, linear identities such as $(\hat{x} + \hat{y}) - \hat{y} = \hat{y}$ or $(3\hat{x}) - \hat{x} = 2\hat{x}$, which are not satisfied in IA, are rigorously verified in AA.

Hence, the application of AA for computing linear functions is characterized by extremely high relative accuracy, which is close to one, resolving the dependency problem affecting IA-based computing. Indeed, in this context, the only endogenous uncertainty source is related to the potential round-off errors, which can be introduced when computing the parameters of the affine form $\hat{z}$. Representing this endogenous uncertainty is not so straightforward in AA as it is in IA, because it is not sufficient to properly amplify the partial deviations z_i. Indeed, this rounding process could induce erroneous correlations between the affine forms sharing the same noise symbols, hence compromising the application of the fundamental invariant of AA. Consequently, in order to model an endogenous uncertainty d affecting the value of partial deviation z_i, preserving the fundamental invariant AA, it is necessary to add a new noise symbol ε_{n+1} and add the term $d\varepsilon_{n+1}$ to $\hat{z}$. In order to make a clear distinction between the noise symbols $\varepsilon_1, \varepsilon_2, \dots, \varepsilon_n$, which model the exogenous uncertainties affecting the value of the uncertain variables, and the new noise symbols added during the computations in order to model the endogenous uncertainty, we named the first as the *primitive noise symbols*.

Let's now consider the application of AA in the task of computing nonlinear functions $z \to f(x, y)$, where x and y are two uncertain variables described by the corresponding affine forms $\hat{x}$ and $\hat{y}$. In this case, it is necessary to identify a mathematical algorithm, which allows us to compute the value of the dependent variable z for any instance of the primitive noise symbols, namely:

$$
\begin{aligned}
z &= f(x_0 + x_1\varepsilon_1 + x_2\varepsilon_2 + \cdots + x_n\varepsilon_n, y_0 + y_1\varepsilon_1 + y_2\varepsilon_2 + \cdots + y_n\varepsilon_n) \\
&= f^*(\varepsilon_1, \varepsilon_2, \dots, \varepsilon_n)
\end{aligned}
\tag{1.70}
$$

where f^* is a nonlinear function from U^n to R, which cannot be expressed by a linear combination of the primitive noise symbols due to the nonlinearity of f. Hence, in order to express the uncertain variable z by an affine form, a proper affine approximation of $f^*(\varepsilon_1, \varepsilon_2, \dots, \varepsilon_n)$ should first be defined over its domain:

$$
f^a(\varepsilon_1, \varepsilon_2, \dots, \varepsilon_n) = z_0 + z_1\varepsilon_1 + z_2\varepsilon_2 + \cdots + z_n\varepsilon_n
\tag{1.71}
$$

Moreover, a new term $z_{n+1}\varepsilon_{n+1}$ should be added to the approximated affine form in order to model the endogenous uncertainty, where $|z_{n+1}|$ is an upper bound of the approximation error, namely:

$$
|z_k| \geq \max\left\{ |f^*(\varepsilon_1, \varepsilon_2, \dots, \varepsilon_n) - f^a(\varepsilon_1, \varepsilon_2, \dots, \varepsilon_n)| : \varepsilon_1, \varepsilon_2, \dots, \varepsilon_n \in U \right\}
\tag{1.72}
$$

and ε_{n+1} is a new noise symbol, which was not defined in any previous computations. Hence, the resulting affine form is described by the following expression:

$$
\begin{aligned}
\hat{z} &= f^a(\varepsilon_1, \varepsilon_2, \dots, \varepsilon_n) + z_{n+1}\varepsilon_{n+1} \\
&= z_0 + z_1\varepsilon_1 + z_2\varepsilon_2 + \cdots + z_n\varepsilon_n + z_{n+1}\varepsilon_{n+1}
\end{aligned}
\tag{1.73}
$$

It is worth noting that this approximation process introduces a new endogenous uncertainty source, which affects the computation accuracy. Indeed, the new noise symbol ε_{n+1} is assumed to describe an independent uncertainty source, which is actually strictly dependent from all the exogenous uncertainties described by the primitive noise symbols $(\varepsilon_1, \varepsilon_2, \dots, \varepsilon_n)$. The lack in modeling this intrinsic correlation is propagated in all the subsequent computations, affecting the precision of the obtained results, which tend to be slightly conservative.

However, a proper choice of the affine approximation f^a makes the error term describing the endogenous uncertainty quadratically dependent from the input affine forms $\hat{x}$ and $\hat{y}$; hence, its magnitude $|z_{n+1}|$ decreases with their ranges. In particular, f^a is frequently expressed by a linear combination of the input affine forms $\hat{x}$ and $\hat{y}$, namely:

$$f^a(\varepsilon_1, \varepsilon_2, \dots, \varepsilon_n) = \alpha\hat{x} + \beta\hat{y} + \zeta \tag{1.74}$$

where α, β, and ζ are unknown parameters. This choice of the affine approximation function represents a proper trade-off between computing accuracy and efficiency for many class of functions. In particular, it could be shown that for "smooth" functions, the error between the affine approximation introduced in (1.74) and the "best" optimal approximation depends quadratically on the ranges of the input affine forms $\hat{x}$ and $\hat{y}$, while for univariate functions, the "best" optimal approximation is that described in (1.74).

Approximating the function f by the affine function described in (1.74) requires the definition of a procedure for computing the unknown parameters α, β, and ζ. The final goal of this identification process is to try and maximize the result accuracy, while lowering the computing complexity and the memory requirements.

An effective and reliable measure of the result accuracy is the magnitude of the partial deviation z_{n+1}, which describes the effect of the endogenous uncertainty, namely the inability of $\hat{z}$ to correctly represent some regions of the true solution domain. Another reliable measure of the result accuracy is the volume of the polytope P_{xyz}, which describes the uncertainty in locating the point (x, y, z) in the domain $\hat{z} = \hat{f}(\hat{x}, \hat{y})$. These accuracy metrics are equivalent after adopting the affine approximation defined in (1.74). In this case, the volume of P_{xyz}, which is a prism with vertical axis and parallel oblique bases, is $2|z_{n+1}|$ times the volume of the joint polytope P_{xy}. Hence, considering that the volume of P_{xyz} does not depend on the affine approximation f^a, the minimization of $|z_{n+1}|$ is equivalent to the minimization of the volume of P_{xyz}.

1.9.3 Chebyshev Approximation of Univariate Nonaffine Functions

An effective way of computing the unknown parameters of the affine approximation functions introduced in (1.74) for a fixed univariate function f is based on the adoption of the Chebyshev approximation theory. This computes the parameters of the affine form which minimize the maximum absolute error over specified domains according to the following theorem and corollary (Pearson, 1990):

Theorem 1.2 *Let $f : I = [a, b] \to R$ be a bounded and continuous function, and h be the affine function approximating f in I and minimizing the maximum absolute error over I. Then, there should exist three points $u, v, w \in I$, where the error $f(x) - h(x)$ is maximum, and its sign alternates on these three points.*

Corollary 1.1 *Let $f : I = [a, b] \to R$ be a bounded and twice differentiable function, whose second derivative conserves its sign in I. Let $f^a(x) = \alpha x + \zeta$ be the Chebyshev affine approximation of f in I. Then:*

- *$\alpha = (f(b) - f(a))/(b - a)$;*
- *The maximum absolute error occurs with the same sign at the extreme points of I, and with the opposite sign $\forall u \in I : f'(u) = \alpha$;*
- *The term ζ satisfies the following equation: $\alpha u + \zeta = (f(u) + r(u))/2$, where $r(x)$ is the line interpolating the points $(a, f(a))$ and $(b, f(b))$;*
- *The maximum absolute error is $\delta = |f(u) - r(u)|/2$.*

Starting from these theoretical results, it is possible to define an effective algorithm for finding the parameters of the affine approximation functions α and ζ, once the equation $f'(u) = \alpha$ is solvable.

Let's consider, for example, the AA-based implementation of the square root function $z = \sqrt{x}$ over the interval $I = [a, b]$ with $a \geq 0$, which requires approximating the following nonaffine function:

$$\sqrt{\hat{x}} = \sqrt{x_0 + x_1 \varepsilon_1 + x_2 \varepsilon_2 + \cdots + x_n \varepsilon_n} \tag{1.75}$$

According to the Chebyshev approximation theory for univariate functions, the affine approximation minimizing the maximum absolute error is:

$$\alpha \hat{x} + \zeta = \alpha(x_0 + x_1 \varepsilon_1 + x_2 \varepsilon_2 + \cdots + x_n \varepsilon_n) + \zeta \tag{1.76}$$

As the second derivative of $\sqrt{x}$ does not change sign over I, the parameters of this affine form can be computed according to the formula derived in Theorem 2. In particular, α can be computed as:

$$\alpha = \frac{\sqrt{b} - \sqrt{a}}{b - a} \tag{1.77}$$

The point u is the solution of the following equation:

$$\frac{1}{2\sqrt{u}} = \alpha \Rightarrow u = \frac{1}{4\alpha^2} \tag{1.78}$$

Moreover, the term ζ can be computed as:

$$\zeta = \frac{f(u) + r(u)}{2} - \alpha u = \frac{\sqrt{a} + \sqrt{b}}{8} + \frac{1}{2} \frac{\sqrt{a}\sqrt{b}}{\sqrt{a} + \sqrt{b}} \tag{1.79}$$

and the corresponding maximum absolute error, which occurs at a, b, and $c = (\sqrt{a} + \sqrt{b})^2/4$, is:

$$\delta = \frac{f(u) - r(u)}{2} = \frac{1}{8} \frac{(\sqrt{b} - \sqrt{a})^2}{\sqrt{a} + \sqrt{b}} \tag{1.80}$$

Hence, the affine form approximating $z = \sqrt{x}$ is:

$$\hat{z} = z_0 + z_1 \varepsilon_1 + z_2 \varepsilon_2 + \cdots + z_n \varepsilon_n + z_{n+1} \varepsilon_{n+1} \tag{1.81}$$

where ε_{n+1} is a new noise symbol modeling the approximation error, and:

$$z_0 = \alpha x_0 + \zeta$$
$$z_i = \alpha x_i \quad \forall i \in [1, n] \tag{1.82}$$
$$z_{n+1} = \delta$$

To further clarify this concept, let's compute the affine form approximating $z = \sqrt{x}$ for $\hat{x} = 4.5 + 1.5\varepsilon_1$, which results in:

$$\hat{z} = 2.1062 + 0.3587\varepsilon_1 + 0.0154\varepsilon_2 \tag{1.83}$$

The geometric interpretation of this affine approximation can be inferred by analyzing Figure 1.3.

Observe that the range of $\hat{z}$ is $[1.7321, 2.4803]$, which is wider than the range computed using IA, namely $\sqrt{[3, 6]} = [1.7320, 2.4495]$.

This overestimation is caused by the effect of the new noise symbol ε_2, which is assumed to be independent from the other noise symbols, although it is correlated to them by a nonlinear dependency. In particular, the term $0.0154\varepsilon_2$, which models the endogenous uncertainty, tends to be negative when the other terms of the affine form tend to the maximum value. Indeed, a rigorous range analysis shows that the maximum value assumed by $\sqrt{x}$ for $x \in [3, 6]$ is $\sqrt{6} = 2.4495$, which corresponds to $\varepsilon_1 = 1$ and $\varepsilon_2 = -1$, namely:

$$z_0 + z_1 - z_2 = 2.1062 + 0.3587 - 0.0154 = 2.4495 \tag{1.84}$$

The independence of the noise symbols requires considering the sum (difference) of the magnitude of all the partial deviations in computing the upper (lower) bound of the corresponding affine form:

$$UB(\hat{z}) = z_0 + z_1 + z_2 = 2.1062 + 0.3587 + 0.0154 = 2.4803$$
$$LB(\hat{z}) = z_0 - z_1 - z_2 = 2.1062 - 0.3587 - 0.0154 = 1.7321 \tag{1.85}$$

This overestimation problem becomes particularly critical when the range of the input variable x includes negative values. In this case, the range of $\sqrt{z}$ also contains negative values, which is an obvious error, as $\sqrt{x}$ is positively defined.

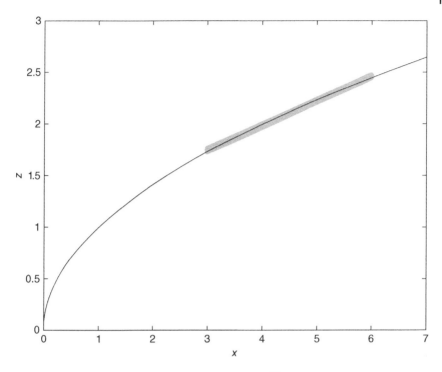

Figure 1.3 Affine approximation of the function $z = \sqrt{x}$.

Hence, the adoption of the Chebyshev approximation theory for approximating nonlinear functions could induce some overestimation effects in computing the corresponding ranges, but it allows reliable modeling of the dependency between the primitive noise symbols, which is important when solving uncertain problems. This feature is even more important when considering that the endogenous uncertainties induced by the approximation errors, which are described by the partial deviations z_{n+1}, depend quadratically on the width of the range of the input variable x. Consequently, when the input ranges decrease, the endogenous uncertainties induced by AA-based operators decrease more quickly compared to the first-order approximation errors characterizing IA.

1.9.4 Multiplication of Affine Forms

A relevant nonaffine function to analyze in the context of AA is the multiplication of affine forms, specifically the affine approximation of the nonlinear function $z = f(x, y) = xy$, where the variables x and y are described by the affine forms $\hat{x}$ and $\hat{y}$, respectively.

The exact expression of this product is a quadratic form of the primitive noise symbols $f^*(\varepsilon_1, \varepsilon_2, \ldots, \varepsilon_n)$, namely:

$$
\begin{aligned}
f^*(\varepsilon_1, \varepsilon_2, \ldots, \varepsilon_n) &= \hat{x}\hat{y} \\
&= \left(x_0 + \sum_{i=1}^{n} x_i \varepsilon_i \right) \left(y_0 + \sum_{i=1}^{n} y_i \varepsilon_i \right) \\
&= x_0 y_0 + \sum_{i=1}^{n} (x_0 y_i + y_0 x_i) \varepsilon_i + \left(\sum_{i=1}^{n} x_i \varepsilon_i \right) \left(\sum_{i=1}^{n} y_i \varepsilon_i \right)
\end{aligned}
\tag{1.86}
$$

This quadratic form can be approximated by the sum of the following affine form:

$$
A(\varepsilon_1, \varepsilon_2, \ldots, \varepsilon_n) = x_0 y_0 + \sum_{i=1}^{n} (x_0 y_i + y_0 x_i) \varepsilon_i
\tag{1.87}
$$

and a proper affine approximation of the following center-symmetric function:

$$
Q(\varepsilon_1, \varepsilon_2, \ldots, \varepsilon_n) = \left(\sum_{i=1}^{n} x_i \varepsilon_i \right) \left(\sum_{i=1}^{n} y_i \varepsilon_i \right) = \sum_{i=1}^{n} \sum_{j=1}^{n} x_i y_j \varepsilon_i \varepsilon_j
\tag{1.88}
$$

that can be approximated by the following center-symmetric affine function:

$$
\frac{Q_M + Q_m}{2} + \frac{Q_M - Q_m}{2} \varepsilon_{n+1}
\tag{1.89}
$$

where ε_{n+1} is a new noise symbol modeling the approximation error, and Q_M and Q_m are the maximum and minimum of Q over U^n, respectively.

Hence, the affine form approximating $\hat{x}\hat{y}$ is:

$$
\hat{z} = A(\varepsilon_1, \varepsilon_2, \ldots, \varepsilon_n) + \frac{Q_M + Q_m}{2} + \frac{Q_M - Q_m}{2} \varepsilon_{n+1}
\tag{1.90}
$$

Unfortunately, the calculation of Q_M and Q_m is not straightforward, since it requires the solution of a combinatorial optimization problem, whose complexity is $O(n \log n)$. Consequently, in order to speed up the computation, the following conservative range estimation could be applied:

$$
-r(\hat{x}) r(\hat{y}) \leq \sum_{i=1}^{n} \sum_{j=1}^{n} x_i y_j \varepsilon_i \varepsilon_j \leq r(\hat{x}) r(\hat{y})
\tag{1.91}
$$

Hence, the affine approximation of the multiplication operator is:

$$
\hat{z} = x_0 y_0 + \sum_{i=1}^{n} (x_0 y_i + y_0 x_i) \varepsilon_i + r(\hat{x}) r(\hat{y}) \varepsilon_{n+1}
\tag{1.92}
$$

It is important to emphasize that this multiplication operator could overestimate the true range of the result (by up to four times). In particular, it is easy to show that the maximum overestimation error could be observed when $\hat{x}$ and $\hat{y}$ are characterized by the same ranges but disjoint noise symbols, as in the following example:

$$\hat{x} = 2 + \varepsilon_1 - \varepsilon_2$$
$$\hat{y} = 1 + \varepsilon_1 + \varepsilon_2 \tag{1.93}$$

In this case, the center-symmetric residual function can be expressed as:

$$Q(\varepsilon_1, \varepsilon_2) = \varepsilon_1^2 + \varepsilon_1\varepsilon_2 - \varepsilon_2\varepsilon_1 - \varepsilon_2^2 = \varepsilon_1^2 - \varepsilon_2^2 \tag{1.94}$$

and its true range is $[-1, 1]$, which differs from the straightforward conservative estimation $r(\hat{x})r(\hat{y}) = [-4, 4]$. In this context, it worth noting that computing the range of the quadratic residual Q using IA-based operators gives the correct value, namely:

$$Q(\varepsilon_1, \varepsilon_2) = [-1, 1]^2 - [-1, 1]^2 = [-1, 1] \tag{1.95}$$

Hence, the use of IA-based operators for computing the range of Q could be a viable solution for enhancing the accuracy of the AA-based multiplication operator.

Despite its over-conservative feature, the straightforward estimation of the range of the quadratic residual defined in (1.91) is frequently adopted in AA-based multiplication, since its overestimation error is quadratic in the widths of the input ranges. Therefore, the corresponding AA-based operator tends to become more accurate, compared to the corresponding IA-based operator, as the input ranges get thinner. To demonstrate this feature, let's compute the following product using the AA-based operator defined in (1.92):

$$z = (5 + 2 * \bar{x} + \bar{y})(5 - 2 * \bar{x} + \bar{h}) \tag{1.96}$$

where $\bar{x} = [-1, 1]$, $\bar{y} = [-1, 1]$, and $\bar{h} = [-2, 2]$. For this, we need to convert these interval variables in the following affine forms:

$$\hat{x} = \varepsilon_1$$
$$\hat{y} = \varepsilon_2 \tag{1.97}$$
$$\hat{h} = 2\varepsilon_3$$

Hence, by applying the previously defined affine operators, it follows that:

$$5 + 2\hat{x} + \hat{y} = 5 + 2\varepsilon_1 + \varepsilon_2$$
$$5 - 2\hat{x} + \hat{h} = 5 - 2\varepsilon_1 + 2\varepsilon_3 \tag{1.98}$$
$$\hat{z} = 25 + 5\varepsilon_2 + 10\varepsilon_3 + (2\varepsilon_1 + \varepsilon_2)(-2\varepsilon_1 + 2\varepsilon_3)$$

Then, considering that the range of the quadratic residual terms can be roughly estimated as $[-12, 12]$, it follows that:

$$\hat{z} = 25 + 5\varepsilon_2 + 10\varepsilon_3 + 12\varepsilon_4 \tag{1.99}$$

And the corresponding range is $[-2, 52]$. However, the true range of the quadratic residual term is $[-12, 2.25]$, and the corresponding range of the uncertain variable z is $[6, 42.5]$. Hence, the relative accuracy of the AA-based multiplication operator is $(42.5 - 6)/(52 + 2) = 0.67$. If we apply the conventional IA-based operators, we obtain the following range:

$$\bar{z} = (5 + 2 * [-1, 1] + [-1, 1])(5 - 2 * [-1, 1] + [-2, 2]) = [2, 72] \qquad (1.100)$$

and the corresponding relative accuracy is $(42.5 - 6)/(72 - 2) = 0.52$. This over-conservative estimation is mainly due to the inability of IA-based computing in modeling the correlations between the uncertain variables. Indeed, IA does not consider that the uncertain variable x is shared by both operands of the multiplication operator, while AA correctly models this dependency by defining two affine forms which share the same noise symbol ε_1.

The multiplication operator defined in (1.92) can be applied to compute the division of affine forms. The main idea is to recast $\hat{x}/\hat{y}$ as $\hat{x}(1/\hat{y})$, which is the product of the first affine form with the reciprocal of the second one. This approach, which requires the application of two nonaffine operations, namely the multiplication and the inverse operator, is characterized by a quadratic convergence, preserving the correlations between $\hat{x}$ and $\hat{y}$.

In order to compute the reciprocal of the affine form $\hat{y}$, it is possible to follow the same approach deployed for deriving the multiplication operator. In particular, let's assume that the range of $\hat{y}$ is $[a, b]$, and it does not include the zero. In this case, the parameters of the following affine approximation (Stolfi and De Figueiredo, 1997):

$$\hat{z} = \frac{1}{\hat{y}} = \alpha\hat{y} + \zeta + \delta\varepsilon_{n+1} \qquad (1.101)$$

can be computed according to the following expressions:

$$\alpha = -\frac{1}{b^2}$$

$$\zeta = sign(a)\frac{\left(\frac{1}{a} + \alpha a\right) - \left(\frac{1}{b} - \alpha b\right)}{2} \qquad (1.102)$$

$$\delta = \frac{\left(\frac{1}{a} - \alpha a\right) - \left(\frac{1}{b} - \alpha b\right)}{2}$$

It is worth observing that the application of the operator defined in (1.102) could give inaccurate results when an extreme point of the input range tends to overlap zero.

1.9.5 Effects of Recursive Solution Schemes

As shown earlier, each application of a nonaffine operator introduces an independent source of endogenous uncertainty, which is modeled by a new noise symbol. Hence, the number of noise symbols could rapidly diverge in the presence of repetitive application of these operators, such as during the application of recursive computing schemes. In order to limit the number of the new noise symbols, it is suggested to "compact" the affine forms by replacing a proper set of endogenous noise symbols $\sum_k z_{n+k} \varepsilon_{n+k}$ with a single equivalent term $z_N \varepsilon_N$, where:

$$z_N = \sum_k |z_{n+k}| \tag{1.103}$$

and ε_N is a new noise symbol. Thanks to this condensing technique, the maximum number of noise symbols can be constrained, but some dependency information characterizing the uncertain variables could be lost. The loss of information is marginal if the partial deviations of the "compacted" noise symbols $|z_{n+k}|$ are small compared to the other partial deviations $|z_i|$.

1.10 Integrating AA and IA

As previously shown, nonaffine operators preserve the dependency information between the uncertain variables, but they introduce approximation errors; these errors could make the variables ranges wider than those computed by applying conventional IA-based operators. Hence, in order to enhance the overall accuracy, it is possible to properly integrate AA and IA-based computing, in order to try and get the best out of the two techniques, namely to use IA for reducing the overestimation of nonaffine operations, and AA for keeping track of the dependency between the uncertain variables. In particular, the main idea is to represent each uncertain variable x by $\hat{\bar{x}}$, which is a combination of an interval variable and an affine form. Hence, the joint range of two variables $\hat{\bar{x}}$ and $\hat{\bar{y}}$ is the intersection between the joint range of the two affine forms $\hat{x}$ and $\hat{y}$ and the hyper-rectangle $\bar{x} \times \bar{y}$.

In order to manage these hybrid variables, proper computing procedures should be defined for performing any mathematical operators $\hat{\bar{z}} \to \hat{\bar{f}}(\hat{\bar{x}}, \hat{\bar{y}})$. To this aim, IA-based operators are deployed in the task of computing $\bar{f}(\bar{x}, \bar{y})$, and identifying the parameters of the affine approximation for f. This affine approximation is then used to compute $\hat{f}(\hat{x}, \hat{y})$. The final result is then obtained by intersecting the obtained ranges, namely:

$$\bar{z} = \bar{f}(\bar{x}, \bar{y}) \cap [f_0(\hat{x}, \hat{y}) - r(\hat{f}(\hat{x}, \hat{y})), f_0(\hat{x}, \hat{y}) + r(\hat{f}(\hat{x}, \hat{y}))] \tag{1.104}$$

This hybrid approach is expected to enhance the accuracy of pure AA and IA-based computing, producing tighter range estimations also when the two separate techniques fail due to error explosion.

2

Uncertain Power Flow Analysis

Power flow (PF) analysis requires the computation of the magnitude and angle (or the rectangular coordinates) of the voltage phasor for each network bus, for a given set of system parameters, e.g. the active and reactive power demanded/generated, assuming a steady-state and balanced power system operation.

Once the voltage phasors at each bus have been computed, it is possible to determine the main variables which characterize steady-state network operations, i.e. the transmission losses, the active and reactive branch PFs, and the generator reactive power outputs.

Hence, PF analysis can be considered as a nonlinear mapping between the following input (output) variables:

- the real and reactive power (voltage phasor magnitude and angle) injected at the PQ buses;
- the real power injected and the voltage phasor magnitude (reactive power injected and voltage phasor angle) at the PV buses;
- the voltage phasor magnitude and angle (the real and reactive power injected) at the reference or slack bus.

This problem can be formalized by the set of nonlinear equations describing the real power balance at the PV and PQ buses, and the reactive power balance at the PQ buses, which can be expressed in polar coordinates as follows:

$$
\begin{aligned}
P_i^{SP} &= V_i \sum_{k=1}^{N} V_k Y_{ik} \cos\left(\delta_i - \delta_k - \theta_{ik}\right) \quad \forall i \in N_P \\
Q_j^{SP} &= V_j \sum_{k=1}^{N} V_k Y_{jk} \sin\left(\delta_j - \delta_k - \theta_{jk}\right) \quad \forall j \in N_Q
\end{aligned}
\tag{2.1}
$$

Interval Methods for Uncertain Power System Analysis, First Edition. Alfredo Vaccaro.

Alternatively, the PF equations can also be expressed in rectangular coordinates as follows:

$$(V_i^{SP})^2 = e_i^2 + f_i^2 \qquad\qquad \forall i \in N_{PV}$$

$$P_i^{SP} = e_i \sum_{k=1}^{N} \left(e_k\, G_{ik} - f_k\, B_{ik} \right) + f_i \sum_{k=1}^{N} \left(f_k\, G_{ik} + e_k\, B_{ik} \right) \qquad \forall i \in N_P$$

$$Q_j^{SP} = f_j \sum_{k=1}^{N} \left(e_k\, G_{jk} - f_k\, B_{jk} \right) - e_i \sum_{k=1}^{N} \left(f_k\, G_{jk} + e_k\, B_{jk} \right) \qquad \forall j \in N_Q$$

(2.2)

where

- N is the number of network buses;
- N_{PV} is the set of the buses for which the bus voltage magnitude is fixed;
- N_P is the set of the buses for which the active power injected into the grid is fixed;
- N_Q is the set of the buses for which the reactive power injected into the grid is fixed;
- P_i^{SP} and Q_j^{SP} are the real and reactive power injections fixed at the ith and jth bus, respectively;
- $V_i \angle \delta_i = e_i + j f_i$ is the ith bus voltage phasor in polar and rectangular coordinates, respectively;
- $Y_{ik} \angle \theta_{ik} = G_{ik} + j B_{ik}$ is the ikth element of the bus admittance matrix in polar and rectangular coordinates, respectively.

The solution of this system of nonlinear equations requires the deployment of numerical algorithms, which are typically based on Newton–Raphson or fast-decoupled methods. These algorithms estimate the PF solutions within a fixed tolerance by approximating the nonlinear balance equations by first-order Taylor series expansion around an initial solution guess, which is numerically solved by iterative schemes and sparse factorization techniques.

More comprehensive PF analysis requires consideration of the limits on some output variables, such as the technical constraints on the reactive power generated at the PV buses required to model the primary-voltage regulators. The simplest approach that can be used to address this issue is based on the so-called bus-type "switching" scheme. This approach converts the PV-buses where the reactive power limits are violated into PQ-buses by fixing the reactive power injected into the grid to the corresponding maximum/minimum value of the violated limit (depending on whether the generator is underexcited or overexcited). After implementing the bus-switching strategy, the balance equations are solved again; hence, obtaining a new PF solution, in which the voltage magnitudes at the switched buses different are from their original set-points.

A more sophisticated method when modeling the primary voltage regulators, the generator reactive power constraints, and the voltage recovery processes, is to formalize the PF problem as a constrained optimization problem using complementarity conditions, as proposed in Pirnia et al. (2013).

PF analysis is the mathematical backbone of many power system operation tools, such as security analysis and contingency assessment. In this context, the PF equations should be repetitively solved for different operation scenario, and the corresponding solutions should be computed in very short time intervals in order to promptly support system operators.

In order to reduce the computational burden of PF analysis, the balance equations formalized in (2.3) can be approximated using the following assumptions:

1. Neglecting the real components of the bus admittance matrix (i.e. $G_{ij} = 0$, and consequently, $\theta_{ij} = \pi/2$ and $\theta_{ii} = -\pi/2$). This assumption implies that the real part of the equivalent circuit representing the power components, and consequently, the active power losses, is neglected.
2. Assuming "small" bus voltage angle differences, such that $\sin(\delta_i - \delta_j) \simeq \delta_i - \delta_j$. This assumption makes it possible to convert the nonlinear balance equations into linear ones; hence, making it possible to obtain their analytical solution.
3. The voltage magnitude at each network bus is fixed, namely $V_i = 1$ p.u., and the bus voltage angle at the slack bus is specified (e.g. $\delta_{ref} = 0$).

Thanks to these assumptions, the nonlinear balance equations formalized in (2.3) can be expressed as

$$
\begin{aligned}
P_i^{SP} &= \sum_{k=1}^{N} B_{ik} \left(\delta_i - \delta_k \right) && \forall i \in N_P \\
Q_j^{SP} &= -\sum_{k=1}^{N} B_{jk} \cos \left(\delta_j - \delta_k \right) && \forall j \in N_Q
\end{aligned}
\tag{2.3}
$$

When analyzing these approximated equations, it is worth observing that the active and reactive balance equations have been decoupled. In particular, the set of $N - 1$ linear algebraic equations describing the active power balance can be solved in order to compute the $N - 1$ voltage angles for all the network buses except the slack.

2.1 Sources of Uncertainties in Power Flow Analysis

The balance equations formalized in Eq. (2.1) are typically referred to as the deterministic PF equations, since all input data are specified by real numbers, which are either derived from a snapshot of the current/expected power system operation or fixed by the analyst on the basis of several hypothesis about the analyzed power system, such as the forecast power generation/demand profiles. Hence, the corresponding PF solutions describe a single power system operation state, which is representative of the limited set of network conditions corresponding to the input data assumptions. Consequently, if the input data are uncertain, many scenarios should be defined and analyzed in order to have a comprehensive representation of all the possible power system operation states. This situation is not infrequent in power system analysis, where many exogenous and endogenous uncertainties sources may significantly affect the input data values. In particular, exogenous uncertainties are induced by model errors, which could affect the equivalent electrical circuits of the power components, or approximation errors, which mainly derive from rounding errors. Endogenous uncertainties are mainly related to the randomness of the injected active and reactive power profiles, which may vary due to

- the volatility of the power generation profiles, which are ruled by complex market dynamics (Verbic and Ca nizares, 2006);
- the increasing number of distributed and dispersed generator units, which could considerably increase the number of power transactions (Verbic and Ca nizares, 2006);
- the complexities of modeling and forecasting the behavior of market operators, which are mainly ruled by unpredictable economic dynamics, introducing large and correlated uncertainties in short-term operation;
- the rising uncertainties characterizing the power demand profiles, which integrate an increasing number of economic-driven flexible loads, and electric vehicles charging stations; and
- the massive integration of renewable power generators, which introduce large and correlated uncertainties in power system operation (Wan and Parsons, 1993).

Since all these data uncertainties can affect the validity of the deterministic PF solutions, reliable solution techniques for modeling and managing the effects of data uncertainties should be integrated in PF analysis. These techniques should support the analyst in characterizing the uncertainties affecting the input data, by determining the corresponding tolerances and propagating the input uncertainties on the tolerances of the PF solutions; hence, providing insight into the level of confidence of the computed solutions. The ability to compute the tolerance of the PF solutions could effectively support the development of sensitivity analysis to parameter variations, which help estimate the rate of change in the PF solution with respect to changes in the input data.

2.2 Solving Uncertain Linearized Power Flow Equations

The solution of the linearized PF problem described in (2.3) in the presence of interval uncertainties requires the solution of the following set of interval linear equations:

$$\bar{\mathbf{P}}^{SP} = \mathbf{B}\,\bar{\delta} \tag{2.4}$$

where $\bar{\mathbf{P}}^{SP}$ is the vector of the uncertain active power injected at the N_P buses, and $\bar{\delta}$ is the corresponding vector of the voltage phasor angle at all the network buses except the slack (where the corresponding angle is assumed as reference and fixed to zero). Note that the uncertainties affecting the active power injections have been described by interval variables; hence, no correlations between the uncertain input variables can be modeled, and for the sake of simplicity, it has been assumed that the elements of the $\mathbf{B}$ matrix are deterministic; hence, neglecting the parameters uncertainties affecting the equivalent circuits of the power components. The last assumption makes the solution of the uncertain linearized PF problem straightforward, since the interval vector describing the uncertain voltage angles can be computed by directly inverting the deterministic matrix $\mathbf{B}$ and applying basic interval-based mathematical operators (e.g. sum and multiplication), namely

$$\bar{\delta} = \mathbf{B}^{-1}\,\bar{\mathbf{P}}^{SP} \tag{2.5}$$

Example 2.1 Let us consider the linearized PF analysis of the three-bus power system depicted in Figure 2.1. This system is characterized by the data summarized in Tables 2.1 and 2.2, which report the bus and the branch data, respectively.

Figure 2.1 Example of a three-bus power system.

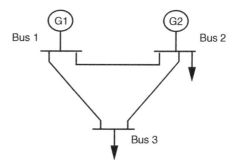

Table 2.1 Bus data.

Bus number	Bus type	P_i^{SP} (MW)
1	Slack	Unknown
2	PV	53
3	PQ	−90

Table 2.2 Line data.

Line number	From bus	To bus	Line reactance (p.u.)^a
1	1	2	0.0576
2	2	3	0.092
3	1	3	0.17

a) 100 MVA base apparent power.

Using these data, the elements of the **B** matrix can be easily computed, resulting in

$$\mathbf{B} = \begin{bmatrix} 23.2435 & -17.3611 & -5.8824 \\ -17.3611 & 28.2307 & -10.8696 \\ -5.8824 & -10.8696 & 16.7591 \end{bmatrix} \tag{2.6}$$

and the deterministic solution of the linearized PF problem can be computed by disregarding the line and column corresponding to the slack bus, in this case the first one, and inverting the corresponding matrix as follows:

$$\delta = \begin{bmatrix} \delta_2 \\ \delta_3 \end{bmatrix} = \begin{bmatrix} 28.2307 & -10.8696 \\ -10.8696 & 16.7591 \end{bmatrix}^{-1} \begin{bmatrix} 0.53 \\ -0.90 \end{bmatrix} = \begin{bmatrix} -0.0025 \\ -0.0554 \end{bmatrix} \tag{2.7}$$

The corresponding active PFs on each line can then be computed as

$$P_{12} = \frac{\delta_1 - \delta_2}{0.0576} = \frac{0.0025}{0.0576} = 0.0434 \text{ p.u. } (4.34 \text{ MW})$$

$$P_{23} = \frac{\delta_2 - \delta_3}{0.092} = \frac{-0.0025 + 0.0554}{0.092} = 0.5750 \text{ p.u. } (57.50 \text{ MW}) \tag{2.8}$$

$$P_{13} = \frac{\delta_1 - \delta_3}{0.17} = \frac{0.0554}{0.17} = 0.3259 \text{ p.u. } (32.59 \text{ MW})$$

Let us now assume that the active power injections are affected by a $\pm 10\%$ uncertainty; hence, they can be described by the following interval variables:

$$\bar{\mathbf{P}}^{SP} = \begin{bmatrix} [0.4770, \ 0.5830] \\ [-0.99, -0.81] \end{bmatrix} \tag{2.9}$$

The corresponding interval variables describing the uncertain voltage angles can be computed as

$$\bar{\delta} = \begin{bmatrix} \bar{\delta}_2 \\ \bar{\delta}_3 \end{bmatrix} = \begin{bmatrix} 28.2307 & -10.8696 \\ -10.8696 & 16.7591 \end{bmatrix}^{-1} \begin{bmatrix} [0.4770, \ 0.5830] \\ [-0.99, -0.81] \end{bmatrix}$$

$$= \begin{bmatrix} [-0.0078, \ 0.0028] \\ [-0.0642, -0.0465] \end{bmatrix} \tag{2.10}$$

and the corresponding interval variables describing the active power flows on each line can be computed as

$$\bar{P}_{12} = \frac{-[-0.0078, \ 0.0028]}{0.0576}$$

$$= [0.0473, \ 0.1354] \text{ p.u. } ([-4.7255, 13.5321]\text{MW})$$

$$\bar{P}_{23} = \frac{[-0.0078, \ 0.0028] - [-0.0642, -0.0465]}{0.092} \tag{2.11}$$

$$= [0.4214, \ 0.7267] \text{ p.u. } ([42.1436, \ 72.6626] \text{ MW})$$

$$\bar{P}_{13} = \frac{-[-0.0642, -0.0465]}{0.17}$$

$$= [0.2739, \ 0.3773] \text{ p.u. } ([27.3921, \ 37.7222] \text{ MW})$$

It is worth observing that on the basis on the fundamental invariance of range analysis, the application of interval arithmetic allows computing reliable enclosures of all the possible values of the voltage angles and the active PFs resulting from the solution of the deterministic linearized PF problem for all the possible combination of the input uncertainties. However, it should be noted that the application of interval analysis (IA) does not model the correlation between the uncertainties affecting the input variables, which are assumed to vary independently over the corresponding ranges.

A more sophisticated approach aimed at modeling uncertain correlations between the input variables in uncertain linearized PF analysis is based on the use of affine arithmetic (AA), which requires expressing the active power injected at the N_P buses by affine forms and computing the corresponding affine forms describing the voltage angle as follows:

$$\hat{\delta}_i = \sum_{k=1}^{N} \tilde{B}_{ik} \hat{P}_k^{SP} = \sum_{k=1}^{N} \tilde{B}_{ik} \left(P_{k0}^{SP} + \sum_{h=1}^{n} P_{kh}^{SP} \epsilon_h \right) \tag{2.12}$$

where $\tilde{B} = B^{-1}$. Note that solving the uncertain linearized PF problem by AA does not introduce endogenous uncertainties (e.g. approximation errors), since all the operators requested for solving the problem are affine operators.

Example 2.2 Let us solve the uncertain linearized PF problem described in Example 2.1, by assuming that the injected active power fixed at the N_p buses ranges over $\pm 10\%$ of their deterministic value due to the effect of the same uncertainty source. In this case, the affine forms describing the power injection share the same noise symbol and can be described as

$$\hat{P}_2^{SP} = 0.53 + 0.053 \, \epsilon_1$$
$$\hat{P}_3^{SP} = -0.9 - 0.09 \, \epsilon_1$$

(2.13)

and the corresponding affine forms describing the voltage angles can be computed as

$$
\begin{bmatrix} \hat{\delta}_2 \\ \hat{\delta}_3 \end{bmatrix} = \begin{bmatrix} 28.2307 & -10.8696 \\ -10.8696 & 16.7591 \end{bmatrix}^{-1} \begin{bmatrix} 0.53 + 0.053 \, \epsilon_1 \\ -0.9 - 0.09 \, \epsilon_1 \end{bmatrix}
$$
$$
= \begin{bmatrix} 0.0472 & 0.0306 \\ 0.0306 & 0.0795 \end{bmatrix} \left(\begin{bmatrix} 0.53 \\ -0.9 \end{bmatrix} + \begin{bmatrix} 0.053 \\ -0.09 \end{bmatrix} \epsilon_1 \right)
$$
$$
= \begin{bmatrix} -0.0026 \\ -0.0554 \end{bmatrix} + \begin{bmatrix} -0.0003 \\ -0.0056 \end{bmatrix} \epsilon_1
$$

(2.14)

The corresponding affine forms describing the active PF on each line can then be computed as follows:

$$\hat{P}_{12} = \frac{-(-0.0026 - 0.0003\epsilon_1)}{0.0576} = 0.0451 + 0.0052\epsilon_1 \text{ p.u.}$$

$$\hat{P}_{23} = \frac{(-0.0026 + 0.0554) + (-0.0003 + 0.0056)\epsilon_1}{0.092} = 0.0527 + 0.0052\epsilon_1 \text{ p.u.}$$

$$\hat{P}_{13} = \frac{0.0554 + 0.0056\epsilon_1}{0.17} = 0.3258 + 0.0329\epsilon_1 \text{ p.u.}$$

(2.15)

Finally, the corresponding bounds of these affine forms are calculated as

$$\bar{P}_{12} = [3.9899, \, 5.0299] \text{ MW}$$
$$\bar{P}_{23} = [4.7499, \, 5.7899] \text{ MW}$$
$$\bar{P}_{13} = [29.2899, \, 35.8699] \text{ MW}$$

(2.16)

This result confirms the flexibility of computing with affine forms. Indeed, if the uncertainty sources affecting the active power injections are independent, then it is possible to compute the corresponding problem solutions by repeating the previously described procedure with the following uncorrelated affine forms:

$$\hat{P}_2^{SP} = 0.53 + 0.053\ \epsilon_1$$
$$\hat{P}_3^{SP} = -0.9 - 0.09\ \epsilon_2$$

(2.17)

The results are summarized here:

$$\begin{bmatrix} \hat{\delta}_2 \\ \hat{\delta}_3 \end{bmatrix} = \begin{bmatrix} 0.0472 & 0.0306 \\ 0.0306 & 0.0795 \end{bmatrix} \left(\begin{bmatrix} 0.53 \\ -0.9 \end{bmatrix} + \begin{bmatrix} 0.053 \\ 0 \end{bmatrix} \epsilon_1 + \begin{bmatrix} 0 \\ -0.09 \end{bmatrix} \epsilon_2 \right)$$

$$= \begin{bmatrix} -0.0026 \\ -0.0554 \end{bmatrix} + \begin{bmatrix} 0.0025 \\ 0.0016 \end{bmatrix} \epsilon_1 + \begin{bmatrix} -0.0028 \\ -0.0072 \end{bmatrix} \epsilon_2$$

(2.18)

$$\hat{P}_{12} = \frac{-(-0.0026 + 0.0025\epsilon_1 - 0.0028\epsilon_2)}{0.0576} = 0.0451 - 0.0435\epsilon_1$$
$$+ 0.0486\epsilon_2 \text{ p.u.}$$

$$\hat{P}_{23} = \frac{(-0.0026 + 0.0554) + (0.0025 - 0.0016)\epsilon_1 + (-0.0028 + 0.0072)\epsilon_2}{0.092}$$
$$= 0.586 + 0.0008\epsilon_1 + 0.0043\epsilon_2 \text{ p.u.}$$

$$\hat{P}_{13} = \frac{0.0554 - 0.0016\epsilon_1 + 0.072\epsilon_2}{0.17} = 0.3176 - 0.0095\epsilon_1 + 0.0423\epsilon_2 \text{ p.u.}$$

(2.19)

With these affine forms, it is possible to promptly and reliably estimate the impact of each single source of data uncertainty on the problem solutions; hence, supporting large-scale sensitivity analysis of the problem solutions to the tolerance of the active power injected into the grid.

If we would also like to consider the effects of circuit parameters uncertainty on the problem solutions, then the problem become more complex, as the matrix **B** is now an interval matrix. In this case, the inversion of B requires the deployment of interval operators for solving system of linear interval equations, such as those described in Section 1.5 or in Rump (2013).

Example 2.3 Let us solve the uncertain linearized PF problem described in Example 2.1, by assuming that both the elements of the matrix **B** and the injected active power fixed at the N_P buses range over $\pm 10\%$ of their deterministic value. In this case, the following system of linear interval equations should be solved:

$$\begin{bmatrix} [25.4076, 31.0538] & [-11.9566, -9.7826] \\ [-11.9566, -9.7826] & [15.0831, 18.4351] \end{bmatrix} \begin{bmatrix} \bar{\delta}_2 \\ \bar{\delta}_3 \end{bmatrix} = \begin{bmatrix} [0.4770, 0.5830] \\ [-0.99, -0.81] \end{bmatrix}$$

(2.20)

This problem can be solved by computing the inverse of the interval matrix **B** as follows:

$$\begin{bmatrix} \bar{\delta}_2 \\ \bar{\delta}_3 \end{bmatrix} = \begin{bmatrix} [0.0316, 0.0628] & [0.0114, 0.0498] \\ [0.0114, 0.0498] & [0.0533, 0.1058] \end{bmatrix} \begin{bmatrix} [0.4770, \ 0.5830] \\ [-0.99, -0.81] \end{bmatrix}$$

$$= \begin{bmatrix} [-0.0359, 0.0308] \\ [-0.1013, -0.0094] \end{bmatrix} \tag{2.21}$$

Which allows computation of the intervals containing all the bus voltage angle values for any combination of the uncertainties affecting both the injected active powers and the electrical network parameters.

2.3 Solving Uncertain Power Flow Equations

The solution of PF equations in the presence of interval uncertainty by naive IA-based operators is not straightforward due to the wrapping effect and the dependency problem, which introduces excessive over-conservativism in the computed solution bounds. This phenomena could be appreciated by considering the following example:

Example 2.4 Let us consider the four-bus power system defined on pp. 337–338 of Grainger and Stevenson (1994), whose bus and branch data are summarized in Tables 2.3 and 2.4, respectively.

Table 2.3 Bus data.

Bus number	Bus type	P_i^{SP} (MW)	Q_j^{SP} (MVAr)	V_i (p.u.)	δ_i (rad)
1	Slack	Unknown	Unknown	1	0
2	PQ	−170	−105.35	Unknown	Unknown
3	PQ	−200	−123.94	Unknown	Unknown
4	PV	238	Unknown	1.02	Unknown

Let us consider both the active and reactive power injected in N_p buses to be affected by ±10% uncorrelated uncertainties. An accurate estimation of the corresponding ranges of the voltage magnitude and angles at the N_p buses can be obtained by applying a Monte Carlo simulation, which repetitively solves the deterministic PF problem for a large number of input data sampled in the

Table 2.4 Line data.

From	To	r (p.u.)	x (p.u.)	b (p.u.)
1	2	0.01008	0.0504	0.1025
1	3	0.00744	0.0372	0.0775
2	4	0.00744	0.0372	0.0775
3	4	0.01272	0.0636	0.1275

corresponding uncertainty ranges. By applying this technique for 1000 samples, the following solution bounds are obtained (Table 2.5):

Table 2.5 Bounds of the PF solution.

$\bar{\delta}_i$ (rad)	$\bar{V}_i$ (p.u.)	$\bar{e}_i$ (p.u.)	$\bar{f}_i$ (p.u.)
0	1	1	0
[−0.0340, 0.0002]	[0.9791, 0.9856]	[0.9788, 0.9856]	[−0.0336, −0.0005]
[−0.0461, −0.0193]	[0.9645, 0.9733]	[0.9637, 0.9729]	[−0.0444, −0.0208]
[0.0023, 0.0514]	1.02	[1.0187, 1.0200]	[0.0014, 0.0511]

Once the interval variables have been defined, we can compute the active and reactive power injected into the network buses by putting these interval variables back into the PF equations; hence, obtaining the following values:

$$\bar{P}_1 = [51.0794, 222.2881] \text{ MW} \qquad \bar{Q}_1 = [-1.6484, 169.5603] \text{ MVAr}$$
$$\bar{P}_2 = [-334.0900, -6.1734] \text{ MW} \qquad \bar{Q}_2 = [-269.5864, 58.3302] \text{ MVAr}$$
$$\bar{P}_3 = [-315.4447, -85.6603] \text{ MW} \qquad \bar{Q}_3 = [-239.1065, -9.3221] \text{ MVAr}$$
$$\bar{P}_4 = [37.5559, 440.1140] \text{ MW} \qquad \bar{Q}_4 = [-69.1917, 333.3665] \text{ MVAr}$$

$$(2.22)$$

As expected, the obtained results include the corresponding fixed intervals $\bar{P}_i^{SP}$ and $\bar{Q}_j^{SP}$, but the computed ranges are so conservative that the obtained solution is not useful at all. As previously discussed, this issue is mainly a consequence of the inability of IA to keep track of correlations between the computed quantities and results in over-estimating the solution bounds. This is visualized in Figure 2.2, where the joint range of two output variables, namely δ_3 and δ_4, sampled by the

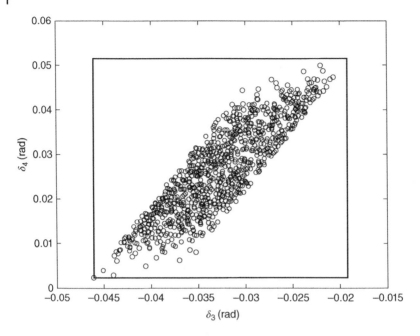

Figure 2.2 Joint range of the power flow solutions.

Monte Carlo simulation are shown. By analyzing this figure, it is worth observing the strong correlation between these variables, which are neglected in IA. Indeed, the joint range of these two variables estimated by IA is represented by the box drawn by the continuous line, which includes large regions which are outside the actual solution space. Hence, more effective reliable computing techniques based on AA will be introduced in Sections 2.3.1 and 2.3.2 to solve this problem.

2.3.1 Optimization-Based Method

When applying AA for uncertain PF problem, the input data, namely the voltage magnitude fixed at the PV buses and the active/reactive power injected at the N_P and N_Q buses, respectively, are represented by the following affine forms:

$$\hat{V}_t^F = V_{t_0}^F$$

$$\hat{P}_i^{SP} = P_{i_0}^F + \sum_{h=1}^{n} P_{i_h}^{SP} \epsilon_h$$

$$\hat{Q}_j^{SP} = Q_{j_0}^F + \sum_{h=1}^{n} Q_{j_h}^{SP} \epsilon_h$$

$$\forall t \in N_{PV}, \ i \in N_P, \ j \in N_Q$$

(2.23)

Observe that, without loss of generality, it has been assumed that the exogenous uncertainties are described by n independent noise symbols, which affect only the active and reactive power injections. Indeed, the typical sources of uncertainties affecting power system operation are those related to the active and reactive power in loads and the active power in generators (specifically, those associated with elastic loads and intermittent sources). The consequence of these uncertainties is the lack of determinism of the voltage phasors at each power system bus, which can be described by the following affine forms:

$$\hat{e}_i = e_{i_0} + \sum_{h=1}^{n} e_{i_h} \epsilon_h$$

$$\hat{f}_i = f_{i_0} + \sum_{h=1}^{n} f_{i_h} \epsilon_h$$

$$\forall i \in N_P$$

The parameters of these affine forms can be determined by solving the following set of nonlinear affine equations, which have been obtained by representing the PF equations formalized in (2.2) using AA-based operators:

$$(V_t^F)^2 = \hat{e}_t^2 + \hat{f}_t^2$$

$$\hat{P}_i^{SP} = \sum_{k=1}^{N} G_{ik}(\hat{e}_i \hat{e}_k + \hat{f}_i \hat{f}_k) + B_{ik}(\hat{f}_i \hat{e}_k - \hat{e}_i \hat{f}_k)$$

$$\hat{Q}_j^{SP} = \sum_{k=1}^{N} G_{jk}(\hat{f}_j \hat{e}_k - \hat{e}_j \hat{f}_k) - B_{jk}(\hat{f}_j \hat{f}_k + \hat{e}_j \hat{e}_k)$$

(2.24)

$$\forall t \in N_{PV}, \quad i \in N_P, \quad j \in N_Q$$

Hence, by using the AA-based multiplication operator defined in (1.92), these equations can be recast as the following set of nonlinear deterministic equations:

$$(V_{t_0}^F)^2 = e_{t_0}^2 + f_{t_0}^2$$

$$P_{i_0}^{SP} = \sum_{k=1}^{N} G_{ik}(e_{i_0} e_{k_0} + f_{i_0} f_{k_0}) + B_{ik}(f_{i_0} e_{k_0} - e_{i_0} f_{k_0})$$

$$P_{i_h}^{SP} = \sum_{k=1}^{N} G_{ik}(e_{i_0} e_{k_h} + e_{i_h} e_{k_0} + f_{i_0} f_{k_h} + f_{i_h} f_{k_0})$$

$$+ B_{ik}(f_{i_0} e_{k_h} - f_{i_h} e_{k_0} - e_{i_0} f_{k_h} - e_{i_h} f_{k_0}) + U_{P_i,h}$$

(2.25)

$$Q_{j_0}^{SP} = \sum_{k=1}^{N} G_{jk}(f_{j_0} e_{k_0} - e_{j_0} f_{k_0}) - B_{jk}(f_{j_0} f_{k_0} + e_{j_0} e_{k_0})$$

$$Q_{j_h}^{SP} = \sum_{k=1}^{N} G_{jk}(f_{j_0} e_{k_h} + f_{j_h} e_{k_0} - f_{j_0} f_{k_h} - f_{j_h} e_{k_0})$$

$$- B_{jk}(f_{j_0} f_{k_h} - f_{j_h} f_{k_0} - e_{j_0} e_{k_h} - e_{j_h} e_{k_0}) + U_{Q_j,h}$$

$$\forall t \in N_{PV}, \quad i \in N_P, \quad j \in N_Q, \quad h \in [1, n]$$

where $U_{P_i,h}$ and $U_{Q_j,h}$ represent the amplification coefficients of the hth partial deviation of the computed active and reactive powers, which model the endogenous uncertainties introduced by the AA-based multiplication operator. In particular, rather than introducing a new noise symbol for each AA-based multiplication in (2.25), the corresponding approximation errors are spread over the partial deviations of the primitive noise symbols, as in the following example:

$$\hat{z} = z_0 + \sum_{h=1}^{n} z_h \epsilon_h + z_{n+1} \epsilon_{n+1} = z_0 + \sum_{h=1}^{n} (z_h + U_{z,h}) \epsilon_h$$

$$U_{z,h} = \frac{|z_h|}{\sum_{k=1}^{n} |z_k|} |z_{n+1}|$$

(2.26)

Observe that this scheme preserves the true range of $\hat{z}$, since

$$\sum_{h=1}^{n} (|z_h| + U_{z,h}) = \sum_{h=1}^{n} \left(|z_h| + \frac{|z_h|}{\sum_{k=1}^{n} |z_k|} |z_{n+1}| \right) = \sum_{h=1}^{n} |z_h| + |z_{n+1}|$$

(2.27)

Hence, each AA-based multiplication in (2.25) is computed as follows:

$$\hat{x}\hat{y} = x_0 y_0 + \sum_{h=1}^{n} (x_0 y_h + y_0 x_h) \epsilon_h + r(\hat{x}) r(\hat{y}) \epsilon_{n+1}$$

$$= x_0 y_0 + \sum_{h=1}^{n} (x_0 y_h + y_0 x_h + U_{z,h}) \epsilon_h$$

(2.28)

$$U_{z,h} = \frac{|x_0 y_h + y_0 x_h|}{\sum_{k=1}^{n} |x_0 y_k + y_0 x_k|} r(\hat{x}) r(\hat{y})$$

This equation expresses the amplification coefficients $U_{P_i,h}$ and $U_{Q_j,h}$ in function of the central values and partial deviations of the unknown affine forms $\hat{e}_i$ and $\hat{f}_i$.

The set of deterministic nonlinear equations formalized in (2.25) is composed of $2N - 1 + (2N - 1 - N_G)n$ equations with as many unknowns. It can be solved by conventional Newton-based numerical methods, whose convergence can be enhanced by computing the sparse Jacobian matrix of the equations (2.25), expressed as

$$
J = \begin{bmatrix}
\dfrac{\partial P_0}{\partial e_0} & \dfrac{\partial P_0}{\partial f_0} & \cdots & \dfrac{\partial P_0}{\partial e_{N_{es}}} & \dfrac{\partial P_0}{\partial f_{N_{es}}} \\[2ex]
\dfrac{\partial Q_0}{\partial e_0} & \dfrac{\partial Q_0}{\partial f_0} & \cdots & \dfrac{\partial Q_0}{\partial e_{N_{es}}} & \dfrac{\partial Q_0}{\partial f_{N_{es}}} \\[2ex]
\vdots & \vdots & \vdots & \vdots & \vdots \\[2ex]
\dfrac{\partial P_{N_{es}}}{\partial e_0} & \dfrac{\partial P_{N_{es}}}{\partial f_0} & \cdots & \dfrac{\partial P_{N_{es}}}{\partial e_{N_{es}}} & \dfrac{\partial P_{N_{es}}}{\partial f_{N_{es}}} \\[2ex]
\dfrac{\partial Q_{N_{es}}}{\partial e_0} & \dfrac{\partial Q_{N_{es}}}{\partial f_0} & \cdots & \dfrac{\partial Q_{N_{es}}}{\partial e_{N_{es}}} & \dfrac{\partial Q_{N_{es}}}{\partial f_{N_{es}}} \\[2ex]
\dfrac{\partial V_{t0}}{\partial e_0} & \dfrac{\partial V_{t0}}{\partial f_0} & \cdots & \dfrac{\partial V_{t0}}{\partial e_{tN_{es}}} & \dfrac{\partial V_{t0}}{\partial f_{N_{es}}} \\[2ex]
\vdots & \vdots & \vdots & \vdots & \vdots \\[2ex]
\dfrac{\partial V_{tN_{es}}}{\partial e_0} & \dfrac{\partial V_{tN_{es}}}{\partial f_0} & \cdots & \dfrac{\partial V_{tN_{es}}}{\partial e_{N_{es}}} & \dfrac{\partial V_{tN_{es}}}{\partial f_{N_{es}}}
\end{bmatrix} \tag{2.29}
$$

Example 2.5 Let us solve the uncertain PF problem defined in Example 2.4 by AA. To achieve this, the corresponding uncertain active/power injections should be modeled by the following affine forms:

$$\hat{P}_2^{SP} = -1.7 + 0.17\epsilon_1$$

$$\hat{Q}_2^{SP} = -1.0535 + 0.10535\epsilon_1$$

$$\hat{P}_3^{SP} = -2 + 0.2\epsilon_2 \tag{2.30}$$

$$\hat{Q}_3^{SP} = -1.2394 + 0.12394\epsilon_2$$

$$\hat{P}_4^{SP} = 2.38 + 0.238\epsilon_3$$

Note that these affine forms share three noise symbols, which aim at modeling the uncertainty sources affecting the active and reactive power injections at the PQ buses, i.e. at buses 2 and 3, and the active power injection at the PV bus, i.e. at bus 4. Moreover, observe that the affine forms describing the active and reactive power injected at the PQ buses share the same noise symbol, e.g. $\hat{P}_2^{SP}$ and $\hat{Q}_2^{SP}$ share ϵ_1, which means that they are correlated.

The corresponding affine forms describing the rectangular coordinates of the voltage phasors at the N_p buses can be computed by solving the set of deterministic nonlinear equations formalized in (2.25), hence obtaining the following results:

$$\hat{e}_1 = 1$$
$$\hat{e}_1 = 0$$
$$\hat{e}_2 = 0.9829 - 0.0065\epsilon_1 - 0.0071\epsilon_2 - 0.0010\epsilon_3$$
$$\hat{f}_2 = -0.0169 + 0.0078\epsilon_1 + 0.0035\epsilon_2 + 0.0066\epsilon_3$$
$$\hat{e}_3 = 0.9689 - 0.0064\epsilon_1 - 0.0001\epsilon_2 - 0.0006\epsilon_3$$
$$\hat{f}_3 = -0.0317 + 0.0029\epsilon_1 + 0.0063\epsilon_2 + 0.0042\epsilon_3 \qquad (2.31)$$
$$\hat{e}_4 = 1.0207 - 0.0167\epsilon_1 - 0.0120\epsilon_2 - 0.0020\epsilon_3$$
$$\hat{f}_4 = 0.0268 + 0.0089\epsilon_1 + 0.0066\epsilon_2 + 0.0120\epsilon_3$$

These solutions are extremely useful in power system operation, since they allow propagating the exogenous uncertainty sources affecting the active/reactive power injections on the bus voltage phasors. The accuracy of the AA-based solution can be assessed by comparing the bounds of the computed affine forms with those obtained by the Monte Carlo simulation, which are reported in Table 2.6.

Table 2.6 Bounds of the PF solution.

Bus	$\bar{e}_i$ (rad)	$\bar{f}_i$ (p.u.)	$\bar{e}_i$ (p.u.)	$\bar{f}_i$ (p.u.)
2	[0.9683, 0.9975]	[−0.0348, 0.0010]	[0.9788, 0.9856]	[−0.0336, −0.0005]
3	[0.9618, 0.9760]	[−0.0451, −0.0184]	[0.9637, 0.9729]	[−0.0444, −0.0208]
4	[0.9900, 1.0515]	[−0.0008, 0.0544]	[1.0187, 1.0200]	[0.0014, 0.0511]

The comparative analysis of these data confirms the effectiveness of AA in computing reliable and accurate enclosures of the PF solutions.

2.3.2 Domain Contraction Method

A different solution scheme that can be adopted to solve uncertain PF problems by AA is based on the domain contraction method proposed in Vaccaro et al. (2010). The main idea is to represent the polar coordinates of the bus voltage phasors by the following affine forms:

$$V_i = V_{i,0} + \sum_{j \in N_P} V_{i,j}^P \epsilon_{P_j} + \sum_{k \in N_Q} V_{i,k}^Q \epsilon_{Q_k} \quad \forall i \in N_Q$$
$$\delta_i = \delta_{i,0} + \sum_{j \in N_P} \delta_{i,j}^P \epsilon_{P_j} + \sum_{k \in N_Q} \delta_{i,k}^Q \epsilon_{Q_k} \quad \forall i \in N_P \qquad (2.32)$$

where ϵ_{P_j} and ϵ_{Q_k} are the noise symbols modeling the uncertainty of the active power injection at the jth bus and the reactive power injection at the kth bus, respectively; $V_{i,0}$ and $\delta_{i,0}$ are the central values of the ith bus voltage magnitude and angle, respectively; $V_{i,j}^P$ and $V_{i,j}^Q$ are the partial deviations of the ith bus voltage magnitude due to the active power injected at the jth bus and due to the reactive power injected at the jth bus, respectively; $\delta_{i,j}^P$ and $\delta_{i,j}^Q$ are the partial deviations of the ith bus voltage angle due to the active power injected at the jth bus and due to the reactive power injected at the jth bus, respectively.

The central values of the affine forms (2.32) can be determined by solving the deterministic PF equations (2.3) fixing all the input data uncertainties to their corresponding central values as follows:

$$
\begin{aligned}
P_i^{SP} &= m\left(\left[P_{i,min}^{SP}, P_{i,max}^{SP}\right]\right) = \frac{P_{i,max}^{SP} - P_{i,min}^{SP}}{2} \quad \forall i \in N_P \\
Q_i^{SP} &= m\left(\left[Q_{i,min}^{SP}, Q_{i,max}^{SP}\right]\right) = \frac{Q_{i,max}^{SP} - Q_{i,min}^{SP}}{2} \quad \forall i \in N_Q
\end{aligned}
\tag{2.33}
$$

Moreover, a first approximated estimation of the partial deviations of the affine forms (2.32) can be obtained by computing the sensitivity coefficients of the bus voltage magnitudes and angles with respect to the uncertain inputs at their central values, i.e.

$$
\begin{aligned}
V_{i,j}^P &= \frac{\partial V_i}{\partial P_j}\Big|_0 \Delta P_j \quad V_{i,k}^Q = \frac{\partial V_i}{\partial Q_k}\Big|_0 \Delta Q_k \quad \forall j \in N_P, \ \forall k, \ i \in N_Q \\
\delta_{i,j}^P &= \frac{\partial \delta_i}{\partial P_j}\Big|_0 \Delta P_j \quad \delta_{i,k}^Q = \frac{\partial \delta_i}{\partial Q_k}\Big|_0 \Delta Q_k \quad \forall i, \ j \in N_P, \ \forall k \in N_Q
\end{aligned}
\tag{2.34}
$$

Observe that for linearized PF problems, which require computing only affine expressions, these affine forms would be the exact AA-based PF solution. However, due to the nonlinearity of the PF equations, the obtained affine forms are usually an underestimation of the solution domain (Grabowski et al., 2008). Hence, as suggested in Grabowski et al. (2008), an outer affine estimation of the PF solutions can be obtained by multiplying each approximated partial deviation by a proper amplification coefficient.

Starting from this initial outer solution, a "domain contraction" method is used to properly narrow its bounds. For this purpose, the amplified partial deviations are adopted to determine the affine forms of the voltage phasors defined in (2.32), which are plugged into the right-hand side of the PF equations (2.3) to determine the following affine forms of the injected active and reactive powers:

$$
\begin{aligned}
\hat{Q}_i &= Q_{i,0} + \sum_{j \in N_P} Q_{i,j}^P \varepsilon_{P_j} + \sum_{k \in N_Q} Q_{i,k}^Q \varepsilon_{Q_k} + \sum_{h \in N_N} Q_{i,h} \varepsilon_h \quad \forall i \in N_Q \\
\hat{P}_i &= P_{i,0} + \sum_{j \in N_P} P_{i,j}^P \varepsilon_{P_j} + \sum_{k \in N_Q} P_{i,k}^Q \varepsilon_{Q_k} + \sum_{h \in N_N} P_{i,h} \varepsilon_h \quad \forall i \in N_P
\end{aligned}
\tag{2.35}
$$

where $\hat{P}_i$ and $\hat{Q}_i$ are the affine forms of the computed active and reactive power injections at the ith bus, respectively; ε_h are new noise symbols modeling the endogenous uncertainties (e.g. approximation errors) generated by the nonaffine operations (N_N is the set of these new noise symbols); $Q_{i,0}$, $Q_{i,j}^P$, $Q_{i,k}^Q$, $P_{i,0}$, $P_{i,j}^P$, and $P_{i,k}^Q$ are the computed central values and the partial deviations of the affine forms of the calculated active and reactive powers injected at the ith node; and $Q_{i,h}$ and $P_{i,h}$ are the partial deviations of the noise symbols ε_h, modeling the approximation errors generated by nonaffine operations.

The AA-based mathematical operators described in Section 1.9.2 and the Chebyshev approximations of the AA-based sinusoidal functions described in Stolfi and De Figueiredo (1997) are deployed for computing $Q_{i,j}^P$, $Q_{i,k}^Q$, $Q_{i,h}$, $P_{i,j}^P$, $P_{i,k}^Q$, and $P_{i,h}$. The obtained affine forms (2.35) can then be expressed by the following matrix formalism:

$$
\begin{bmatrix} \hat{Q}_1 \\ \cdots \\ \hat{Q}_{n_Q} \\ \hat{P}_1 \\ \cdots \\ \hat{P}_{n_P} \end{bmatrix} = \begin{bmatrix} Q_{1,0} \\ \cdots \\ Q_{n_Q,0} \\ P_{1,0} \\ \cdots \\ P_{n_P,0} \end{bmatrix} + \begin{bmatrix} Q_{1,1}^P & \cdots & Q_{1,n_P}^P & Q_{1,1}^Q & \cdots & Q_{1,n_Q}^Q \\ \cdots & \cdots & \cdots & \cdots & \cdots & \cdots \\ Q_{n_Q,1}^P & \cdots & Q_{n_Q,n_P}^P & Q_{n_Q,1}^Q & \cdots & Q_{n_Q,n_Q}^Q \\ P_{1,1}^P & \cdots & P_{1,n_P}^P & P_{1,1}^Q & \cdots & P_{1,n_Q}^Q \\ \cdots & \cdots & \cdots & \cdots & \cdots & \cdots \\ P_{n_P,1}^P & \cdots & P_{n_P,n_P}^P & P_{n_P,1}^Q & \cdots & P_{n_P,n_Q}^Q \end{bmatrix} \begin{bmatrix} \varepsilon_{P_1} \\ \cdots \\ \varepsilon_{P_{n_P}} \\ \varepsilon_{Q_1} \\ \cdots \\ \varepsilon_{Q_{n_Q}} \end{bmatrix}
$$

$$
+ \begin{bmatrix} Q_{1,1} & \cdots & Q_{1,n_N} \\ \cdots & \cdots & \cdots \\ Q_{n_Q,1} & \cdots & Q_{n_Q,n_N} \\ P_{1,1} & \cdots & P_{1,n_N} \\ \cdots & \cdots & \cdots \\ P_{n_P,1} & \cdots & P_{n_P,n_N} \end{bmatrix} \begin{bmatrix} \varepsilon_1 \\ \cdots \\ \cdots \\ \cdots \\ \cdots \\ \varepsilon_{n_N} \end{bmatrix} \tag{2.36}
$$

where n_P and n_Q represent the number of *PV* nodes and *PQ* nodes and n_N is the number of the noise symbols modeling the endogenous uncertainties. More generally, (2.36) can be written as

$$f(X) = AX + B \tag{2.37}$$

where

$$
A = \begin{bmatrix} Q_{1,1}^P & \cdots & Q_{1,n_P}^P & Q_{1,1}^Q & \cdots & Q_{1,n_Q}^Q \\ \cdots & \cdots & \cdots & \cdots & \cdots & \cdots \\ Q_{n_Q,1}^P & \cdots & Q_{n_Q,n_P}^P & Q_{n_Q,1}^Q & \cdots & Q_{n_Q,n_Q}^Q \\ P_{1,1}^P & \cdots & P_{1,n_P}^P & P_{1,1}^Q & \cdots & P_{1,n_Q}^Q \\ \cdots & \cdots & \cdots & \cdots & \cdots & \cdots \\ P_{n_P,1}^P & \cdots & P_{n_P,n_P}^P & P_{n_P,1}^Q & \cdots & P_{n_P,n_Q}^Q \end{bmatrix} \tag{2.38}
$$

$$
X = \begin{bmatrix} \varepsilon_{P_1} \\ \cdots \\ \varepsilon_{P_{n_P}} \\ \varepsilon_{Q_1} \\ \cdots \\ \varepsilon_{Q_{n_Q}} \end{bmatrix} \tag{2.39}
$$

$$
B = \begin{bmatrix} Q_{1,0} \\ \cdots \\ Q_{n_Q,0} \\ P_{1,0} \\ \cdots \\ P_{n_P,0} \end{bmatrix} + \begin{bmatrix} Q_{1,1} & \cdots & Q_{1,n_N} \\ \cdots & \cdots & \cdots \\ Q_{n_Q,1} & \cdots & Q_{n_Q,n_N} \\ P_{1,1} & \cdots & P_{1,n_N} \\ \cdots & \cdots & \cdots \\ P_{n_P,1} & \cdots & P_{n_P,n_N} \end{bmatrix} \begin{bmatrix} \varepsilon_1 \\ \cdots \\ \cdots \\ \cdots \\ \cdots \\ \varepsilon_{n_N} \end{bmatrix} \tag{2.40}
$$

Note that, A is a deterministic real-valued matrix; X is the vector that should be contracted, with initial values for each of its components set at $[-1, 1]$; and B is an interval vector describing the endogenous uncertainties, which cannot be contracted because it models the effect of approximation errors introduced into the computational process by the nonaffine operators. A reliable enclosure of the PF solutions can then be obtained by contracting the vector X so that

$$
AX + B = f^{SP} \tag{2.41}
$$

where f^{SP} is the interval vector describing the range of the fixed active and reactive powers injected at the N_P and N_Q buses, respectively:

$$
f^{SP} = \begin{bmatrix} \left[Q_{1,min}^{SP}, Q_{1,max}^{SP} \right] \\ \cdots \\ [Q_{n_Q,min}^{SP}, Q_{n_Q,max}^{SP}] \\ [P_{1,min}^{SP}, P_{1,max}^{SP}] \\ \cdots \\ [P_{n_P,min}^{SP}, P_{n_P,max}^{SP}] \end{bmatrix} \tag{2.42}
$$

Hence, the overall problem has been reduced to the solution of the following set of interval linear equations:

$$
AX = f^{SP} - B = C \tag{2.43}
$$

where A is a fixed real-value matrix.

The problem formalized in (2.43) can be recast as the following $n_P + n_Q$ constrained linear optimization problems:

$$\min \quad (\varepsilon_{Q_k}, \varepsilon_{P_j}) \quad \forall k \in N_Q, \quad \forall j \in N_P$$

$$\text{s.t.} \quad -1 \le \varepsilon_{Q_k} \le 1, \quad -1 \le \varepsilon_{P_j} \le 1$$

$$inf(C_i) \le \sum_{j \in N_P} A_{ij}\varepsilon_{P_j} + \sum_{k \in N_Q} A_{ik}\varepsilon_{Q_k} \le sup(C_i) \; \forall i \in [1, N_P + N_Q]$$

(2.44)

$$\max \quad (\varepsilon_{Q_k}, \varepsilon_{P_j}) \quad \forall k \in N_Q, \quad \forall j \in N_P$$

$$\text{s.t.} \quad -1 \le \varepsilon_{Q_k} \le 1, \quad -1 \le \varepsilon_{P_j} \le 1$$

$$inf(C_i) \le \sum_{j \in N_P} A_{ij}\varepsilon_{P_j} + \sum_{k \in N_Q} A_{ik}\varepsilon_{Q_k} \le sup(C_i) \; \forall i \in [1, N_P + N_Q]$$

(2.45)

These deterministic linear programming problems can be effectively solved by using conventional mathematical tools (Dolan and Moré, 2002). The corresponding PF solution is then obtained as

$$V_i = V_{i,0} + \sum_{j \in N_P} V_{i,j}^P [\varepsilon_{P_{j,min}}, \varepsilon_{P_{j,max}}] + \sum_{k \in N_Q} V_{i,k}^Q [\varepsilon_{Q_{k,min}}, \varepsilon_{Q_{k,max}}] \quad \forall i \in N_Q$$

$$\delta_i = \delta_{i,0} + \sum_{j \in N_P} \delta_{i,j}^P [\varepsilon_{P_{j,min}}, \varepsilon_{P_{j,max}}] + \sum_{k \in N_Q} \delta_{i,k}^Q [\varepsilon_{Q_{k,min}}, \varepsilon_{Q_{k,max}}] \quad \forall i \in N_P$$

(2.46)

The described AA-based solution method is a viable alternative to the conventional linearization-based approaches frequently adopted to solve nonlinear interval equations, which require solving the following interval linear problem:

$$F(x_0 + \Delta x) \in F(x) + J(x_0)\Delta x \quad \forall x \in x_0$$

(2.47)

where x_0 is a vector of intervals, J is the interval extension of the Jacobian matrix, and $F(x)$ is a real vector defined by x, which is typically the midpoint of x_0.

The solution of the interval problem formalized in (2.47) requires the inversion of the interval matrix $J(x_0)$, which is a nontrivial problem (Kolev and Nenov, 2000; Kolev, 1997), and, as pointed out in Alvarado and Wang (1993) and Barboza et al. (2004), this is one of the main limitations of deploying IA for solving uncertain PF problems. On the other hand, the AA-based problem formalized in (2.43) does not require any interval matrix inversion, making the solution process flexible, scalable, and computationally efficient.

Example 2.6 Let us apply the described AA-based solution technique to solve the uncertain PF problem described in Example 2.4. For this, we need to first solve the deterministic PF problem, i.e. by fixing all the input data uncertainties to their corresponding central values, obtaining the data reported in Table 2.7.

Table 2.7 Deterministic PF solution.

Bus	$\mathring{V}_{i,0}$ (p.u.)	$\delta_{i,0}$ (rad)
1	1.0000	0
2	0.9824	−0.0170
3	0.9690	−0.0327
4	1.0200	0.0266

Then, a first estimation of the partial deviations of the affine forms describing the PF solutions can be obtained by computing the sensitivity coefficients of the voltage magnitudes and angles, as reported in Tables 2.8 and 2.9, respectively.

Table 2.8 Sensitivity coefficients of the bus voltage magnitudes.

Bus	$V_{i,1}$ (p.u.)	$V_{i,2}$ (p.u.)	$V_{i,3}$ (p.u.)
1	0	0	0
2	0.003238	−0.000023	0.000028
3	−0.000029	0.004357	0.000028
4	0	0	0

Table 2.9 Sensitivity coefficients of the bus voltage angles.

Bus	$\delta_{i,1}$ (rad)	$\delta_{i,2}$ (rad)	$\delta_{i,3}$ (rad)
1	0	0	0
2	0.0062	0.0022	−0.0023
3	0.0018	0.0058	−0.0015
4	0.0049	0.0037	−0.0039

Table 2.10 Bounds of the PF solution.

Bus	$\bar{V}_i$ (rad)	$\bar{\delta}_i$ (rad)
1	[1.0000, 1.0000]	[0, 0]
2	[0.9791, 0.9857]	[−0.0380, 0.0039]
3	[0.9646, 0.9734]	[−0.0507, −0.0147]
4	[1.0200, 1.0200]	[0.0018, 0.0514]

Then, by amplifying these partial deviations by a factor of 2 and solving 5 constrained linear optimization problems, the following affine forms describing the PF solutions can be computed:

$$\hat{V}_1 = 1.02$$
$$\hat{\delta}_1 = 0$$
$$\hat{V}_2 = 0.9824 - 0.0032\varepsilon_1$$
$$\hat{\delta}_2 = -0.0170\varepsilon_1 - 0.0122\varepsilon_2 - 0.0043\varepsilon_2 + 0.0045\varepsilon_3$$
$$\hat{V}_3 = 0.9690 - 0.0044\varepsilon_2$$
$$\hat{\delta}_3 = -0.0036\varepsilon_1 - 0.0115\varepsilon_2 + 0.0029\varepsilon_3$$
$$\hat{V}_4 = 1.0200$$
$$\hat{\delta}_4 = 0.0266 - 0.0097\varepsilon_1 - 0.0073\varepsilon_2 + 0.0078\varepsilon_3$$

(2.48)

The corresponding bounds, reported in Table 2.10, confirm the robustness of the computed AA-based PF solutions.

3

Uncertain Optimal Power Flow Analysis

Optimal power flow (OPF) analysis aims at computing the power system operation state, which satisfies specific economic, environmental, or technical criteria without violating a set of constraints describing grid and power equipment operating limits. The solution of this problem requires the identification of the optimal configuration of a proper vector of decision variables u that minimizes one or more cost functions f_i, subject to a number of nonlinear equality g_j and inequality constraints h_k. Hence, this problem can be formalized by the following constrained nonlinear multiobjective programming problem:

$$
\begin{aligned}
&\min_{(x,u)} \quad f_i(x, u) \qquad \forall i \in [1, p] \\
&\text{s.t.} \quad g_j(x, u) = 0 \quad \forall j \in [1, n] \\
&\qquad\quad h_k(x, u) < 0 \quad \forall k \in [1, m]
\end{aligned}
\tag{3.1}
$$

where x is the vector of the state variables, p is the number of scalar cost functions, and n and m are the number of equality and inequality constraints, respectively. This programming problem can also be expressed in a vectorial form as follows:

$$
\begin{aligned}
&\min_{(x,u)} \quad f(x, u) \\
&\text{s.t.} \quad g(x, u) = 0 \\
&\qquad\quad h(x, u) < 0
\end{aligned}
\tag{3.2}
$$

where $f(.)$ is the p-dimensional cost function vector, and $g(.)$ and $h(.)$ are the n-dimensional and m-dimensional vectors, representing the equality and inequality constraints, respectively.

The decision variables in (3.2) depend on the particular power system application and could include both real-valued and integer variables, such as the active power generated by dispatchable generators (i.e. optimal power dispatch), the set points of the primary voltage regulators (i.e. reactive power dispatch), the loading factor (i.e. voltage stability analysis), and the set of the available generators (i.e.

Interval Methods for Uncertain Power System Analysis, First Edition. Alfredo Vaccaro.
© 2023 The Institute of Electrical and Electronics Engineers, Inc. Published 2023 by John Wiley & Sons, Inc.

unit commitment). Hence, the OPF is an instance of nonconvex mixed integer/ nonlinear programming (MINLP) problems.

The power system state variables include the components of the voltage phasor at PQ buses, the voltage phasor angle and the reactive power injected at the PV buses, and the active and reactive power injected at the slack bus. The inequality constraints include the maximum allowable power flows for the power lines, the limits of the decision variables, i.e. $u_{min,i} \leq u_i \leq u_{max,i}$, $\forall i \in [1, n_u]$, and for some state variables, i.e. $x_{min,i} \leq x_i \leq x_{max,i}$, $\forall i \in [1, n_x]$. Moreover, the decision and the state variables should satisfy the power flow (PF) equations (2.3), which represent the equality constraints for (3.1) and (3.2).

The cost functions $f(.)$ could be defined according to technical, environmental, and economic criteria, such as the generation production costs, the power system losses, the average voltage deviation, and the greenhouse gas emissions. Since these design objectives are frequently competing, and considering the nonconvexity and nonlinear characteristics of the PF equations, the OPF problem has no unique solution and a proper trade-off between the costs functions should be identified. In this context, an effective solution strategy is based on the deployment of the weighted global criterion method, which makes it possible to combine all cost functions into a single utility function as follows:

$$U(x, u) = \sum_{i=1}^{p} \left(\omega_i f_i(x, u) \right)^r f_i(x, u) \geq 0 \tag{3.3}$$

where the weights ω_i, so that $\sum_{i=1}^{p} \omega_i = 1$, $\omega_i > 0$, should be properly defined by the analyst according to the relevance of the cost functions.

A large number of programming algorithms have been proposed for solving OPF problems, e.g. nonlinear programming (Shoults and Sun, 1982), quadratic programming (Momoh, 1989; Burchett et al., 1982), and linear programming (Pandya and Joshi, 2008). Many methods aim at formalizing and solving the set of nonlinear equations describing the Karush–Kuhn–Tucker optimality conditions. These methods, which are based on iterative Newton-based solution schemes, can effectively handle both equality and inequality constraints, with the latter being modeled by adding quadratic penalty terms to the cost function (Sun et al., 1984). Another solution scheme widely adopted for solving OPF problems is based on the Interior Point method (Momoh and Zhu, 1999), which converts the inequality constraints into equivalent equalities by defining proper nonnegative slack variables. A self-concordant barrier function of these slack variables is then integrated into the cost function and multiplied by a barrier parameter, which gradually tends to zero during the solution process. Another method that does not require the application of heuristic schemes for reducing the barrier parameter is based on the unlimited point algorithm (Tognola and Bacher, 1999). This method converts the optimality conditions into a set of nonlinear equations by implementing a proper transformation of the slack and dual variables of the inequality constraints.

The problem formalized in (3.1) is referred to as deterministic OPF because all input data are described by deterministic variables, which could be derived from a snapshot of power system operation or could be defined according to proper assumptions about the analyzed power system. Hence, the solution of this deterministic optimization problem refers to a particular power system state, which is representative of a limited set of power system operation conditions. Thus, when the input data are affected by data uncertainties, a large number of system operation scenarios should be defined and analyzed. The source of uncertainties affecting the input data of OPF problems, which are the same as those described in Section 2.1, could affect the problem solutions to a considerable extent. Hence, the deployment of reliable solution techniques for estimating both the input data and the solution tolerance therefore providing insight into the level of confidence of OPF solutions are required. Moreover, reliable methodologies for uncertain OPF analyses could support large-scale sensitivity analysis of input data variations, which can estimate the rate of change in the OPF solution with respect to changes in the input data.

3.1 Range Analysis-Based Solution

The integration of affine arithmetic (AA) and range analysis can be applied to solve OPF problems in the presence of uncertain input data modeled by real compact intervals, as proposed in Vaccaro et al. (2013). The main idea is to formalize the uncertain OPF problem by the following nonlinear interval optimization problem:

$$\min_{\bar{z}} \quad \bar{f}(\bar{z})$$

$$\text{s.t.} \quad \bar{g}_j(\bar{z}) = 0 \quad \forall j \in [1, n] \tag{3.4}$$
$$\bar{h}_k(\bar{z}) < 0 \quad \forall k \in [1, m]$$

where the bounds of the interval extension of the cost function $\bar{f}(\bar{z})$, the n equality constrained functions $\bar{g}_j$, and the m inequality constrained functions $\bar{h}_k$, are defined as follows:

$$\bar{f}(\bar{z}) = [f_{low}(\bar{z}), f_{up}(\bar{z})] \tag{3.5}$$

$$\bar{g}_j(\bar{z}) = [g_{j,low}(\bar{z}), g_{j,up}(\bar{z})] \quad \forall j \in [1, n] \tag{3.6}$$

$$\bar{h}_k(\bar{z}) = [h_{k,low}(\bar{z}), h_{k,up}(\bar{z})] \quad \forall k \in [1, m] \tag{3.7}$$

where $f_{low}(\bar{z})$, $g_{j,low}(\bar{z})$ $\forall j \in [1, n]$ and $h_{k,low}(\bar{z})$ $\forall k \in [1, m]$ are the lower boundary functions, while $f_{up}(\bar{z})$, $g_{j,up}(\bar{z})$ $\forall j \in [1, n]$ and $h_{k,up}(\bar{z})$ $\forall k \in [1, m]$ are the corresponding upper boundary functions. If these bounds are computed by applying AA-based operators, then the inclusion isotonicity property of AA makes it possible to solve the uncertain optimization problem formalized in (3.4) by the following theorem (Levin, 1999):

Theorem 3.1 *(Fundamental Theorem of Range Analysis)* *For the interval function $\hat{\Gamma}(\hat{\chi}) = [\Gamma_{low}(\hat{\chi}), \Gamma_{up}(\hat{\chi})]$ to take the minimum (maximum) value at χ^* in its domain G, it is necessary and sufficient that the boundary functions $\Gamma_{low}(\hat{\chi})$ and $\Gamma_{up}(\hat{\chi})$ take the minimum (maximum) value at the same point:*

$$\hat{\Gamma}(\chi^*) = [\Gamma_{low}(\chi^*), \Gamma_{up}(\chi^*)] = \min_{\hat{\chi} \in G} \quad \hat{\Gamma}(\hat{\chi}) = [\Gamma_{low}(\hat{\chi}), \Gamma_{up}(\hat{\chi})]$$

$$\Leftrightarrow \begin{cases} \Gamma_{low}(\chi^*) = \min_{\hat{\chi} \in G} \quad \Gamma_{low}(\hat{\chi}) \\ \Gamma_{up}(\chi^*) = \min_{\hat{\chi} \in G} \quad \Gamma_{up}(\hat{\chi}) \end{cases}$$

$$\hat{\Gamma}(\chi^*) = [\Gamma_{low}(\chi^*), \Gamma_{up}(\chi^*)] = \max_{\hat{\chi} \in G} \quad \hat{\Gamma}(\hat{\chi}) = [\Gamma_{low}(\hat{\chi}), \Gamma_{up}(\hat{\chi})]$$

$$\Leftrightarrow \begin{cases} \Gamma_{low}(\chi^*) = \max_{\hat{\chi} \in G} \quad \Gamma_{low}(\hat{\chi}) \\ \Gamma_{up}(\chi^*) = \max_{\hat{\chi} \in G} \quad \Gamma_{up}(\hat{\chi}) \end{cases}$$

The most important consequence of this theorem is that it allows the bounds of an interval function in a given domain to be computed by determining the extreme points of its lower- and upper-boundary functions in the same domain. A direct consequence of this result is that the solution bounds of the interval problem formalized in (3.4) can be computed by solving two deterministic problems, namely the lower- and upper-boundary problems (Hladík, 2011; Lemke et al., 2002), which can be formalized by the following optimization problems (Levin, 2004; Hladík, 2011; Lemke et al., 2002):

$$\min_{\hat{z}} \quad f_{low}(\hat{z})$$

$$\text{s.t.} \quad g_{j,low}(\hat{z}) = 0 \quad \forall j \in [1, n] \tag{3.8}$$
$$h_{k,up}(\hat{z}) < 0 \quad \forall k \in [1, m]$$

$$\min_{\hat{z}} \quad f_{up}(\hat{z})$$

$$\text{s.t.} \quad g_{j,up}(\hat{z}) = 0 \quad \forall j \in [1, n] \tag{3.9}$$
$$h_{k,up}(\hat{z}) < 0 \quad \forall k \in [1, m]$$

The solution of these programming problems determines the lower and upper bounds of the interval extension of the cost function $\hat{f}(\hat{z}) = [f_{low}(\hat{z}), f_{up}(\hat{z})]$ of the OPF problem (3.4), with the corresponding interval extension of the constraint functions, which are described by the upper and lower bound of the equality constraints $\hat{g}_j(\hat{z}) = [g_{j,low}(\hat{z}), g_{j,up}(\hat{z})] \, \forall j \in [1, n]$, and from the upper bound of the inequality constraints $\hat{h}_k(\hat{z}) = [h_{k,low}(\hat{z}), h_{k,up}(\hat{z})] \, \forall k \in [1, m]$.

Hence, as proposed in Vaccaro et al. (2013), the bounds of the OPF solution can be determined according to the following solution scheme:

1. Determine an outer estimation of the uncertain OPF problem solution, which can be obtained by using the sensitivity-based approach described in (2.34):

$$\hat{z}_{outer} = z_0 + z_1 \varepsilon_1 + \cdots + z_p \varepsilon_p \tag{3.10}$$

2. Solve the lower boundary problem formalized in (3.8) using a conventional deterministic programming tool, obtaining a solution $\varepsilon_{i,low} \, \forall i = [1, p]$.
3. Solve the upper boundary problem formalized in (3.9) using the same programming tool as in Step 1, obtaining a solution $\varepsilon_{i,up} \, \forall i = [1, p]$.
4. Compute the solution bounds as

$$\hat{z} = z_0 + z_1 \varepsilon_{1,opt} + \cdots + z_p \varepsilon_{p,opt}, \tag{3.11}$$

where

$$\varepsilon_{i,opt} = [-\varepsilon_{i,up}, \varepsilon_{i,up}] \cap [-\varepsilon_{i,low}, \varepsilon_{i,low}] \quad \forall i = [1, p] \tag{3.12}$$

3.1.1 Optimal Economic Dispatch

Economic dispatch analysis aims at computing the active power generated by a set of controllable generating units that meet the expected system load at the lowest possible generation cost, while assuring secure and reliable power system operation. The overall problem can be formalized by the following constrained nonlinear optimization-programming problem (Gomez-Exposito et al., 2009):

$$
\begin{aligned}
\min_{(P_1, \ldots, P_{ng})} \quad & \sum_{i=1}^{ng} (\alpha_i + \beta_i P_i + \gamma_i P_i^2) \\
\text{s.t.} \quad & \sum_{i=1}^{ng} P_i = P_D + P_{loss}(P_1, \ldots, P_{ng}) \\
& P_{i,min} \le P_i \le P_{i,max} \qquad \forall i \in [1, ng]
\end{aligned} \tag{3.13}
$$

where P_D is the net active power demand, namely, the difference between the total power demand and the power generated by renewable power generators; n_g is the number of controllable generators; P_i is the power generated by the ith unit;

$P_{i,min}$ and $P_{i,max}$ are the minimum and maximum generation limits of the ith unit, respectively; α_i β_i and γ_i are the cost coefficients of the ith unit; and $P_{loss}(P_1, \ldots, P_{ng})$ are the active power losses that can be computed by using the following simplified equation:

$$P_{loss}(P_1, \ldots, P_{ng}) = \sum_{i=1}^{ng} B_i P_i^2 \tag{3.14}$$

where B_i are fixed loss coefficients.

The lack of determinism in the input data, e.g. due to the uncertainty affecting the power demand, makes this problem a particular instance of the uncertain OPF problem formalized in (3.4), whose solution can be obtained by deploying the solution scheme described in (3.8) and (3.9). For this scheme, the uncertain power demand should be expressed by the following affine form:

$$\hat{P}_D = P_{D,0} + P_{D,1}\varepsilon_1 + \cdots + P_{i,p}\varepsilon_p \tag{3.15}$$

where p is the number of independent uncertainty sources affecting the power demand, while $P_{D,0}$ and $P_{D,1}$ are the corresponding central values and partial deviations, which represent the input data of the optimization problem.

The corresponding active power generated by the controllable generation units is treated as intervals as follows:

$$\hat{P}_i = P_{i,0} + P_{i,1}\varepsilon_1 + \cdots + P_{i,p}\varepsilon_p \quad \forall i \in [1, n_g] \tag{3.16}$$

which yields the following upper and lower bounds:

$$\hat{P}_{i,up} = P_{i,0} + P_{i,1}\varepsilon_{1,up} + \cdots + P_{i,p}\varepsilon_{p,up} \quad \forall i = [1, n_g]$$
$$\hat{P}_{i,low} = P_{i,0} + P_{i,1}\varepsilon_{1,low} + \cdots + P_{i,p}\varepsilon_{p,low} \quad \forall i = [1, n_g] \tag{3.17}$$

where the noise symbols bounds $\varepsilon_{j,up}$ and $\varepsilon_{j,low}$, $\forall j = [1, p]$, can be obtained by solving the following deterministic OPF problems:

$$\min_{(\varepsilon_{1,up}, \ldots, \varepsilon_{p,up})} \quad UB\left(\sum_{i=1}^{n_g}(\alpha_i + \beta_i \hat{P}_{i,up} + \gamma_i \hat{P}_{i,up}^2)\right)$$

$$\text{s.t.} \quad \sum_{i=1}^{n_g} P_{i,0} + \sum_{i=1}^{n_g}\sum_{j=1}^{p}\left|P_{i,j}\,\varepsilon_{j,up}\right|$$

$$= UB\left(P_{D,max} + P_{loss}(\hat{P}_{1,up}, \ldots, \hat{P}_{n_g,up})\right) \tag{3.18}$$

$$P_{i,0} + \sum_{j=1}^{p}\left|P_{i,j}\,\varepsilon_{j,up}\right| \le P_{i,max} \quad \forall i \in [1, n_g]$$

$$P_{i,min} \le P_{i,0} - \sum_{j=1}^{p}\left|P_{i,j}\,\varepsilon_{j,up}\right| \quad \forall i \in [1, n_g]$$

$$\min_{(\epsilon_{1,low},\ldots,\epsilon_{p,low})} \quad LB\left(\sum_{i=1}^{n_g}(\alpha_i + \beta_i \hat{P}_{i,low} + \gamma_i \hat{P}_{i,low}^2)\right)$$

$$\text{s.t.} \quad \sum_{i=1}^{n_g} P_{i,0} - \sum_{i=1}^{n_g}\sum_{j=1}^{p}\left|P_{i,j}\,\epsilon_{j,low}\right|$$

$$= LB\left(P_{D,min} + P_{loss}(\hat{P}_{1,low}, \ldots, \hat{P}_{n_{ga},low})\right) \tag{3.19}$$

$$P_{i,0} + \sum_{j=1}^{p}\left|P_{i,j}\,\epsilon_{j,low}\right| \le P_{i,max} \quad \forall i \in [1,n_g]$$

$$P_{i,min} \le P_{i,0} - \sum_{j=1}^{p}\left|P_{i,j}\,\epsilon_{j,low}\right| \quad \forall i \in [1,n_g]$$

where UB and LB are the upper and lower bound operators for affine forms, which are defined as follows:

$$UB(\hat{x}) = x_0 + r(\hat{x})$$
$$LB(\hat{x}) = x_0 - r(\hat{x}) \tag{3.20}$$

Thus, the upper and lower bound of the affine function describing the production costs can be computed as

$$UB(\hat{f}) = \sum_{k=1}^{ng}\alpha_k + \beta_k P_{k,0} + \gamma_k P_{k,0}^2 + \sum_{h=1}^{p}|\beta_k(P_{k,h} + 2P_{k,0}P_{k,h})| + r(\hat{P}_k)^2$$

$$LB(\hat{f}) = \sum_{k=1}^{ng}\alpha_k + \beta_k P_{k,0} + \gamma_k P_{k,0}^2 - \sum_{h=1}^{p}|\beta_k(P_{k,h} + 2P_{k,0}P_{k,h})| - r(\hat{P}_k)^2 \tag{3.21}$$

Example 3.1 Let us solve the optimal economic dispatch problem for a three-generator power system, which is characterized by the following cost coefficients and operating limits (Loia and Vaccaro, 2014):

$$C_1(P_1) = 500 + 5.3P_1 + 0.004P_1^2 \quad 0 \le P_1 \le 450 \text{ MW}$$
$$C_2(P_2) = 400 + 5.5P_2 + 0.006P_2^2 \quad 0 \le P_2 \le 350 \text{ MW} \tag{3.22}$$
$$C_3(P_3) = 200 + 5.8P_3 + 0.009P_3^2 \quad 0 \le P_3 \le 225 \text{ MW}$$

The active power losses are neglected, and the uncertain net load is described by the following affine form:

$$\hat{P}_D = P_{D0} + P_{D1}\epsilon_1 + P_{D2}\epsilon_2 = 700 + 70\epsilon_1 + 35\epsilon_2 \tag{3.23}$$

where the noise symbols ϵ_1 and ϵ_2 model two independent sources of uncertainty affecting the net load, such as the errors in forecasting renewable generators production and load demand. Consequently, the sum of the production costs, which is the cost function of the uncertain optimization problem, can be expressed as

$$f(\hat{P}_1, \hat{P}_2, \hat{P}_3) = \sum_{k=1}^{3} \alpha_k + \beta_k P_{k,0} + \gamma_k P_{k,0}^2 + \sum_{h=1}^{2} \beta_k (P_{k,h} + 2P_{k,0}P_{k,h})\varepsilon_h + r(\hat{P}_k)^2 \varepsilon_3$$

(3.24)

To compute the upper and lower bounds of the problem solution let us apply the solution algorithm discussed in Section 3.1.1, which requires solving the deterministic optimization problems formalized in (3.18) and (3.19). The obtained solution bounds are summarized in Table 3.1, which also reports the corresponding bounds computed by applying a Monte Carlo-based approach with 5000 simulations.

Table 3.1 Solution bounds of the optimal economic dispatch problem.

$\bar{P}1$ (MW)	$\bar{P}2$ (MW)	$\bar{P}3$ (MW)
[303.08, 401.84]	[185.39, 251.23]	[106.92, 150.82]
[302.50, 402.76]	[159.34, 277.50]	[32.89, 225.00]

The obtained results confirm the robustness of the solution bounds computed by the described approach.

3.1.2 Reactive Power Dispatch

Another interesting example of OPF analysis is the reactive power dispatch, which aims at identifying the optimal set-point of the primary voltage controllers, the active power injected into the reference bus, and the reactive power generated at the N_{PV} buses, given the active and reactive power injected in the N_P and N_Q buses, respectively. The overall problem can be formalized by the following constrained nonlinear programming problem (Gomez-Exposito et al., 2009):

$$\min_{(V_i \ \forall i \in [1,N], \delta_i \ \forall i \in N_P, Q_i \ \forall i \in N_{PV})} \frac{1}{N} \sum_{i=1}^{N} (V_i - 1)^2$$

$$\text{s.t.} \quad P_i^{SP} - V_i \sum_{j=1}^{N} V_j Y_{ij} \cos \left(\delta_i - \delta_j - \theta_{ij} \right) = 0 \quad \forall i \in N_P$$

$$Q_j^{SP} - V_j \sum_{k=1}^{N} V_k Y_{jk} \sin \left(\delta_j - \delta_k - \theta_{jk} \right) = 0 \quad \forall j \in N_Q$$

$$Q_i - V_i \sum_{k=1}^{N} V_k Y_{ik} \sin \left(\delta_i - \delta_k - \theta_{ik} \right) = 0 \quad \forall i \in N_{PV}$$

$$V_{i,min} \le V_i \le V_{i,max} \quad \forall i \in [1, N]$$

$$Q_{i,min} \le Q_i \le Q_{i,max} \quad \forall i \in N_{PV}$$

(3.25)

The solution paradigm formalized in (3.8) and (3.9) can be utilized to efficiently solve this problem in the presence of interval uncertainty. To achieve this, all the state and decision variables of the problem (3.25) are modeled by intervals as follows:

$$
\begin{aligned}
\hat{V}_i &= V_{i,0} + V_{i,1}\varepsilon_1 + \cdots + V_{i,p}\varepsilon_p \quad &\forall i \in [1, N] \\
\hat{\delta}_i &= \delta_{i,0} + \delta_{i,1}\varepsilon_1 + \cdots + \delta_{i,p}\varepsilon_p \quad &\forall i \in N_P \\
\hat{Q}_i &= Q_{i,0} + Q_{i,1}\varepsilon_1 + \cdots + Q_{i,p}\varepsilon_p \quad &\forall i \in N_{PV}
\end{aligned}
\tag{3.26}
$$

which yields the following upper and lower bounds:

$$
\begin{aligned}
\hat{V}_{i,low} &= V_{i,0} + V_{i,1}\varepsilon_{1,low} + \cdots + V_{i,p}\varepsilon_{p,low} \quad &\forall i = [1, N] \\
\hat{V}_{i,up} &= V_{i,0} + V_{i,1}\varepsilon_{1,up} + \cdots + V_{i,p}\varepsilon_{p,up} \quad &\forall i = [1, N] \\
\hat{\delta}_{i,low} &= \delta_{i,0} + \delta_{i,1}\varepsilon_{1,low} + \cdots + \delta_{i,p}\varepsilon_{p,low} \quad &\forall i \in N_P \\
\hat{\delta}_{i,up} &= \delta_{i,0} + \delta_{i,1}\varepsilon_{1,up} + \cdots + \delta_{i,p}\varepsilon_{p,up} \quad &\forall i \in N_P \\
\hat{Q}_{i,low} &= Q_{i,0} + Q_{i,1}\varepsilon_{1,low} + \cdots + Q_{i,p}\varepsilon_{p,low} \quad &\forall i \in N_{PV} \\
\hat{Q}_{i,up} &= Q_{i,0} + Q_{i,1}\varepsilon_{1,up} + \cdots + Q_{i,p}\varepsilon_{p,up} \quad &\forall i \in N_{PV}
\end{aligned}
\tag{3.27}
$$

where the noise symbols bounds $\varepsilon_{j,up}$ and $\varepsilon_{j,low}$ $\forall j = [1, p]$ can be obtained by solving the following deterministic OPF problems:

$$
\min_{(\varepsilon_{1,up},\ldots,\varepsilon_{p,up})} \quad UB\left(\frac{1}{N}\sum_{i=1}^{N}(\hat{V}_{i,up} - 1)^2\right)
$$

$$
\text{s.t.} \quad UB\left(P_i^{SP} - \hat{V}_{i,up}\sum_{j=1}^{N}\hat{V}_{j,up}Y_{ij}\cos\left(\hat{\delta}_{i,up} - \hat{\delta}_{j,up} - \theta_{ij}\right)\right) = 0 \quad \forall i \in N_P
$$

$$
UB\left(Q_j^{SP} - \hat{V}_{j,up}\sum_{k=1}^{N}\hat{V}_{k,up}Y_{jk}\sin\left(\hat{\delta}_{j,up} - \hat{\delta}_{k,up} - \theta_{jk}\right)\right) = 0 \quad \forall j \in N_Q
$$

$$
UB\left(Q_i - \hat{V}_{i,up}\sum_{k=1}^{N}\hat{V}_{k,up}Y_{ik}\sin\left(\hat{\delta}_{i,up} - \hat{\delta}_{k,up} - \theta_{ik}\right)\right) = 0 \quad \forall i \in N_{PV}
$$

$$
V_{i,0} + \sum_{j=1}^{p}\left|V_{i,j}\,\varepsilon_{j,up}\right| \leq V_{i,max} \quad \forall i \in [1, N]
$$

$$
V_{i,min} \leq V_{i,0} + \sum_{j=1}^{p}\left|V_{i,j}\,\varepsilon_{j,up}\right| \quad \forall i \in [1, N]
$$

$$
Q_{i,0} + \sum_{j=1}^{p}\left|Q_{i,j}\,\varepsilon_{j,up}\right| \leq Q_{i,max} \quad \forall i \in N_{PV}
$$

$$
Q_{i,min} \leq Q_{i,0} + \sum_{j=1}^{p}\left|Q_{i,j}\,\varepsilon_{j,up}\right| \quad \forall i \in N_{PV}
$$

$$
\tag{3.28}
$$

$$\min_{(\varepsilon_{1,low},\dots,\varepsilon_{p,low})} \quad LB\left(\frac{1}{N}\sum_{i=1}^{N}(\hat{V}_{i,up}-1)^2\right)$$

$$\text{s.t.} \quad LB\left(P_i^{SP}-\hat{V}_{i,low}\sum_{j=1}^{N}\hat{V}_{j,low}Y_{ij}\cos\left(\hat{\delta}_{i,low}-\hat{\delta}_{j,low}-\theta_{ij}\right)\right)=0 \quad \forall i \in N_P$$

$$LB\left(Q_j^{SP}-\hat{V}_{j,low}\sum_{k=1}^{N}\hat{V}_{k,low}Y_{jk}\sin\left(\hat{\delta}_{j,low}-\hat{\delta}_{k,low}-\theta_{jk}\right)\right)=0 \quad \forall j \in N_Q$$

$$LB\left(Q_i-\hat{V}_{i,low}\sum_{k=1}^{N}\hat{V}_{k,low}Y_{ik}\sin\left(\hat{\delta}_{i,low}-\hat{\delta}_{k,low}-\theta_{ik}\right)\right)=0 \quad \forall i \in N_{PV}$$

$$V_{i,0}-\sum_{j=1}^{p}\left|V_{i,j}\,\varepsilon_{j,low}\right| \le V_{i,max} \qquad\qquad\qquad \forall i \in [1,N]$$

$$V_{i,min} \le V_{i,0}-\sum_{j=1}^{p}\left|V_{i,j}\,\varepsilon_{j,low}\right| \qquad\qquad\qquad \forall i \in [1,N]$$

$$Q_{i,0}-\sum_{j=1}^{p}\left|Q_{i,j}\,\varepsilon_{j,low}\right| \le Q_{i,max} \qquad\qquad\qquad \forall i \in N_{PV}$$

$$Q_{i,min} \le Q_{i,0}-\sum_{j=1}^{p}\left|Q_{i,j}\,\varepsilon_{j,low}\right| \qquad\qquad\qquad \forall i \in N_{PV}$$

$$(3.29)$$

Example 3.2 Let us consider the reactive power dispatch problem for the IEEE 118-bus test system, which represents a portion of the Midwestern American Electric Power System composed of 54 voltage controllable generators, 186 lines, and 64 load buses (Christie, n.d.). In order to test the effectiveness of the described method, a $\pm 20\%$ tolerance for all the specified active and reactive powers have been assumed. The decision variables of the optimization problem are the set-points of the voltage regulators at the PV buses, while the state variables are the reactive power at the PV buses, the voltage magnitude at the PQ buses, and the voltage angle at all the N_P buses. The voltage magnitudes at each bus are constrained to lie in the following allowable range:

$$0.95 \le V_i \le 1.05 \quad \forall i \in [1,118] \qquad\qquad\qquad (3.30)$$

The uncertain decision and state variables of the optimization problem are modeled by the following affine forms:

$$\hat{V}_i = V_{i,0}+\sum_{k=1}^{117}V_{i,k}\varepsilon_k \quad \forall i \in [1,N]$$

$$\hat{\delta}_i = \delta_{i,0}+\sum_{k=1}^{117}\delta_{i,k}\varepsilon_k \quad \forall i \in N_P \qquad\qquad\qquad (3.31)$$

$$\hat{Q}_i = Q_{i,0}+\sum_{k=1}^{117}Q_{i,k}\varepsilon_k \quad \forall i \in N_{PV}$$

where the initial outer estimation of the corresponding central values and the partial deviations was obtained by using the same sensitivity-based approach described in (2.34). The obtained results are summarized in Figures 3.1 and 3.2.

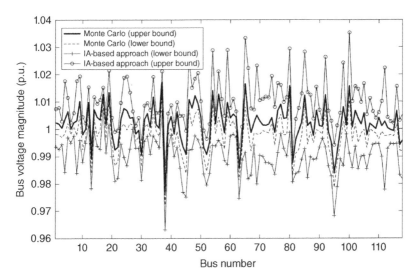

Figure 3.1 Solution bounds – bus voltage magnitudes.

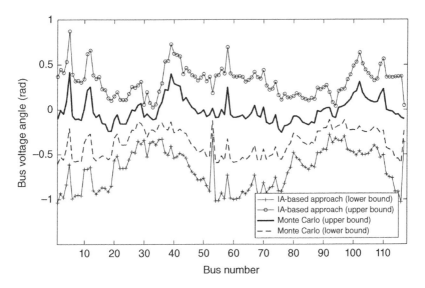

Figure 3.2 Solutions bounds – bus voltage angle.

The analysis of results obtained in this example shows how the described approach can compute reliable approximations of the problem solution bounds when compared to the benchmark bounds estimated using a Monte Carlo-based approach. As expected, the computed bounds are slightly conservative, which is due to the fact that the described AA-based technique estimates the "worst case" combination of the input data uncertainties; this can be considered as an advantage of using reliable computing techniques when solving uncertain optimization problems. Indeed, as discussed in the literature, probabilistic methods that deal with nonprobabilistic uncertainty could underestimate the true "worst case" combination of the data uncertainties, neglecting solutions characterized by very low probability of occurrence. As far as the computational requirements are concerned, it should be noted that the described technique is significantly cheaper than the Monte Carlo approach, since it only requires the solution of two deterministic optimization problems, as opposed to repetitively computing a large number of problem solutions.

3.2 AA-Based Solution

A different computing technique for solving uncertain optimization problems, which are entirely based on AA-based processing, is described here. The main idea is to model the uncertain input data by the following (fixed) affine forms:

$$\hat{P}_j^F = P_{j_0}^F + \sum_{h=1}^n P_{j_h}^F \epsilon_h$$

$$\hat{Q}_i^F = Q_{i_0}^F + \sum_{h=1}^n Q_{i_h}^F \epsilon_h \tag{3.32}$$

$$\forall j \in N_P, \quad i \in N_Q$$

Consequently, the state variables of the OPF problem are represented by the following unknown affine forms:

$$\hat{e}_j = e_{j_0} + \sum_{h=1}^n e_{j_h} \epsilon_h$$

$$\hat{f}_j = f_{j_0} + \sum_{h=1}^n f_{j_h} \epsilon_h \tag{3.33}$$

$$\hat{Q}_t = Q_{t_0} + \sum_{h=1}^n Q_{t_h} \epsilon_h$$

$$\forall j \in N_P, \quad \forall t \in N_{PV}$$

While, without loss of generality, the decision variables have been assumed to be the setpoint of the primary voltage regulators, which are described by the following deterministic variables:

$$V_t \quad \forall t \in N_{PV} \tag{3.34}$$

Consequently, the voltage phasor components at the N_{PV} buses are deterministic variables, and the corresponding partial deviations are zero:

$$e_{t_h} = 0$$
$$f_{t_h} = 0$$
$$\forall t \in N_{PV}$$

(3.35)

Therefore, the solution of this uncertain optimization problem requires computing the deterministic variables defined in (3.35) and the parameters of the affine forms (3.33), which minimize a fixed cost function f_c while satisfying both the equality and inequality technical constraints, thus assuring secure and reliable system operation. The overall problem can be formalized in the AA domain by the following constrained optimization problem:

$$\min_{V_t, \hat{Q}_t, \hat{P}_r, \hat{e}_j, \hat{f}_j} \hat{f}_c$$

$$(V_t)^2 = \hat{e}_t^2 + \hat{f}_{t_0}^2$$

$$\hat{P}_r = \sum_{k=1}^{N} G_{rk}(\hat{e}_r \hat{e}_k + \hat{f}_r \hat{f}_k) + B_{rk}(\hat{f}_r \hat{e}_k - \hat{e}_r \hat{f}_k)$$

$$\hat{P}_j^F = \sum_{k=1}^{N} G_{jk}(\hat{e}_j \hat{e}_k + \hat{f}_j \hat{f}_k) + B_{jk}(\hat{f}_j \hat{e}_k - \hat{e}_j \hat{f}_k)$$

$$\hat{Q}_j^F = \sum_{k=1}^{N} G_{jk}(\hat{f}_j \hat{e}_k - \hat{e}_j \hat{f}_k) - B_{jk}(\hat{f}_j \hat{f}_k + \hat{e}_j \hat{e}_k)$$

$$\hat{Q}_t = \sum_{k=1}^{N} G_{tk}(\hat{f}_t \hat{e}_k - \hat{e}_t \hat{f}_k) - B_{tk}(\hat{f}_t \hat{f}_k + \hat{e}_t \hat{e}_k)$$

$$\hat{Q}_t \le Q_{t_{max}}$$

$$\hat{Q}_{t_0} \ge Q_{t_{min}}$$

$$V_t \le V_{max}$$

$$V_t \ge V_{min}$$

$$\hat{e}_i^2 + \hat{f}_i^2 \le V_{max}^2$$

$$\hat{e}_i^2 + \hat{f}_i^2 \ge V_{min}^2$$

$$\forall t \in N_{PV}, \quad j \in N_p, \quad i \in N_Q, \quad h \in [1, n]$$

(3.36)

This AA-based optimization problem can be recast as the following multiobjective programming problem (Vaccaro and Ca nizares, 2017):

$$\min_{V_t, Q_{t_0}, Q_{t_h}, P_{r_0}, P_{r_h}, e_{j_0}, f_{j_0}, e_{j_h}, f_{j_h}} \left(f_{c_0}, \sum_{h=1}^{n} \left| f_{c_h} \right| \right)$$

$$(V_t)^2 = e_{t_0}^2 + f_{t_0}^2$$

$$P_{r_0} = \sum_{k=1}^{N} G_{rk}(e_{r_0} e_{k_0} + f_{r_0} f_{k_0}) + B_{rk}(f_{r_0} e_{k_0} - e_{r_0} f_{k_0})$$

$$P_{r_h} = \sum_{k=1}^{N} G_{rk}(e_{r_0} e_{k_h} + e_{r_h} e_{k_0} + f_{r_0} f_{k_h} + f_{r_h} f_{k_0})$$
$$\qquad + B_{rk}(f_{r_0} e_{k_h} - f_{r_h} e_{k_0} - e_{r_0} f_{k_h} - e_{r_h} f_{k_0})$$

$$P_{j_0}^F = \sum_{k=1}^{N} G_{jk}(e_{j_0} e_{k_0} + f_{j_0} f_{k_0}) + B_{jk}(f_{j_0} e_{k_0} - e_{j_0} f_{k_0})$$

$$P_{j_h}^F = \sum_{k=1}^{N} G_{jk}(e_{j_0} e_{k_h} + e_{j_h} e_{k_0} + f_{j_0} f_{k_h} + f_{j_h} f_{k_0})$$
$$\qquad + B_{jk}(f_{j_0} e_{k_h} - f_{j_h} e_{k_0} - e_{j_0} f_{k_h} - e_{j_h} f_{k_0})$$

$$Q_{j_0}^F = \sum_{k=1}^{N} G_{jk}(f_{j_0} e_{k_0} - e_{j_0} f_{k_0}) - B_{jk}(f_{j_0} f_{k_0} + e_{j_0} e_{k_0})$$

$$Q_{j_h}^F = \sum_{k=1}^{N} G_{jk}(f_{j_0} e_{k_h} + f_{j_h} e_{k_0} - f_{j_0} f_{k_h} - f_{j_h} e_{k_0})$$
$$\qquad - B_{jk}(f_{j_0} f_{k_h} - f_{j_h} f_{k_0} - e_{j_0} e_{k_h} - e_{j_h} e_{k_0})$$

$$Q_{t_0} = \sum_{k=1}^{N} G_{tk}(f_{t_0} e_{k_0} - e_{t_0} f_{k_0}) - B_{tk}(f_{t_0} f_{k_0} + e_{t_0} e_{k_0})$$

$$Q_{t_h} = \sum_{k=1}^{N} G_{tk}(f_{t_0} e_{k_h} + f_{t_h} e_{k_0} - f_{t_0} f_{k_h} - f_{t_h} e_{k_0})$$
$$\qquad - B_{tk}(f_{t_0} f_{k_h} - f_{t_h} f_{k_0} - e_{t_0} e_{k_h} - e_{t_h} e_{k_0})$$

$$Q_{t_0} + \sum_{h=1}^{n} \left| Q_{t_h} \right| \leq Q_{t_{max}}$$

$$Q_{t_0} - \sum_{h=1}^{n} \left| Q_{t_h} \right| \geq Q_{t_{min}}$$

(3.37)

$$V_t \leq V_{max}$$
$$V_t \geq V_{min}$$

$$e_{i_0}^2 + f_{i_0}^2 + \sum_{h=1}^{n} 2 \left| e_{i_0} e_{i_h} + f_{i_0} f_{i_h} \right| \leq V_{max}^2$$

$$e_{i_0}^2 + f_{i_0}^2 - \sum_{h=1}^{n} 2 \left| e_{i_0} e_{i_h} + f_{i_0} f_{i_h} \right| \geq V_{min}^2$$

$$\forall t \in N_{PV}, \quad j \in N_P, \quad i \in N_Q, \quad h \in [1, n]$$

(3.38)

This formalization is consistent with the AA-based robust optimization approach proposed in Liu et al. (2007), and with risk analysis theory, since the minimization of the affine central values aims to compute the problem solutions without considering the effects of the input data uncertainty, while the minimization of the affine radius aims to compute the problem solutions that exhibit the lowest tolerance to input data uncertainty. A proper trade-off between these two conflicting objectives should be defined by the analyst (achieved with respect to their specific risk strategy). Hence, the minimization of an affine function requires identifying an affine form that minimizes both its central value and its radius. The solution of this deterministic problem makes it possible to compute the parameters of the affine forms describing the state variables and the deterministic values of the decision variables, which satisfy all the equality and inequality constraints for each combination of the input data uncertainties. Hence, the obtained solution minimizes both the central value and the radius of the affine cost function and guarantees that the power system technical constraints are satisfied even for the worst-case scenario, e.g. bus voltage magnitudes are within the acceptable range, reactive power injections are compatible with the generator technical limits, and apparent power flows in each line are less than the maximum allowable transportation limits. Moreover, the formalization of the problem equations by affine forms also models the correlations between the source of uncertainties affecting the input data, which is extremely useful in mitigating the over-conservativism of the computed solution bounds.

Example 3.3   Let us apply the described AA-based technique to solve the uncertain reactive power dispatch problem described in Example 3.2. The obtained results are summarized in Figures 3.3 and 3.4, which represent the computed

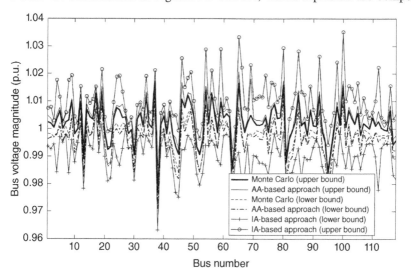

Figure 3.3   Solution bounds – bus voltage magnitude.

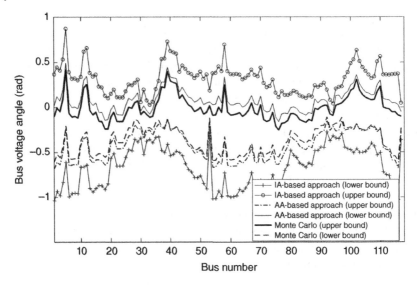

Figure 3.4 Solution bounds – bus voltage angle.

bounds of the bus voltage magnitudes and angles, respectively. When analyzing these figures, it is worth noting that, compared to the range-analysis based approach, the AA-based method can compute more accurate enclosures of the problem solution bounds. However, it should be pointed out that the AA-based method requires larger computational burden.

4

Uncertain Markov Chain Analysis

Markov Chain-based models are the mathematical backbone of many power system tools, ranging from security analysis, to reliability and resilience assessment. Indeed, these models are currently deployed in the task of detecting critical operating states of renewable generators (Zhu et al., 2019); forecasting of power demand and electricity market price profiles (Stephen et al., 2015; González et al., 2005); modeling cascading failures (Wang et al., 2018); identifying the size and location of distributed storage systems (Grillo et al., 2015); and supporting predictive reliability and resilience analysis (Liu et al., 2016; Zhu and Zhang, 2018).

In all these applications, the state transition probabilities of the Markov model, which rule its dynamics, are described by deterministic real-valued parameters. This assumption, which is conventionally assumed in many application domains, might not be suitable in the context of modern power systems, where many complex and correlated data uncertainties could significantly affect the determinism of the data adopted for estimating both the parameters and the outputs of the Markov model. Moreover, other relevant sources of uncertainties that could affect the dynamics of Markov models derive from the estimation of the exogenous variables, which is typically based on statistical analysis (Aien et al., 2016). Since these uncertainties affect the determinism of the state transition probabilities (Brokish and Kirtley, 2009), reliable computing techniques should be applied to model these uncertainties and determine the corresponding bounds of the uncertain state probability trajectories (Blanc and Den Hertog, 2008).

One of the most popular methods for uncertainty propagation analysis of Markov Chain models is based on the adoption of sampling methods, which account for the volatility and stochastic nature of the uncertain data processed for estimating the state transition probabilities. The deployment of this method requires generating a large number of combinations of the uncertain input data and, for each of combination, performing a deterministic model simulation

Interval Methods for Uncertain Power System Analysis, First Edition. Alfredo Vaccaro.
© 2023 The Institute of Electrical and Electronics Engineers, Inc. Published 2023 by John Wiley & Sons, Inc.

(Klauenberg and Elster, 2016). Since the number of input data combinations could be very large, the computational complexity of these solution techniques could be prohibitively high.

To overcome this limitation, nonprobabilistic techniques have recently been proposed for solving Markov Chain models in the presence of data uncertainty. These solution approaches are frequently deployed for solving uncertain problems when data uncertainties stem from incomplete or partial knowledge about the model equations (Smets, 1997), or only imprecise estimations of the model parameters can be determined.

For example, solar radiation can be measured at specific locations, and modeling its spatial distribution in large geographical area using probabilities is extremely complex. Moreover, weather forecast services usually provide only qualitative information about the actual and expected profiles of the environmental variables, which cannot be accurately modeled in a probabilistic form. Thus, the development of computing tools aimed at representing and processing nonprobabilistic knowledge is extremely useful for uncertainty management in Markov chain-based models.

In this context, the application of fuzzy sets has been proposed in Pardo and de la Fuente (2010), where the uncertain state transition probabilities are represented by fuzzy numbers, allowing the computation of the corresponding steady-state probabilities by solving conventional linear programming problems.

Another interesting solution approach is based on the deployment of interval-valued finite Markov Chains (Kozine and Utkin, 2002), which model the state transition probabilities by interval variables (Sen et al., 2006). Consequently, the transition matrix is no longer real-valued but an interval-valued matrix, and the bounds of the steady-state probabilities can be determined by solving an interval system of equations using the techniques described in Chapter 1. However, the deployment of conventional interval analysis (IA)-based operators for solving this problem could be hindered by the wrapping effect and dependency problem, which could add spurious values at each step of the computing chain, resulting in over-conservative bound estimations (Stolfi and De Figueiredo, 1997).

This phenomenon could become a severe limiting factor for the estimation of the bounds of the transient state probabilities, which require the solution of the set of uncertain ordinary differential equations describing the Markov Chain model dynamics.

To address this problem, affine arithmetic (AA) could be deployed to model the uncertainties affecting the state transition probabilities by affine forms and computing the corresponding trajectories of the system state probabilities, as proposed in Pepiciello et al. (2022). The main idea is to apply AA-based operators in order to compute an outer estimation of the transient state probabilities bounds, considering the correlations between the input data uncertainties.

This AA-based Markov Chain model can be deployed in the same application domains as conventional Markov Chains, but with the added advantage of modeling the uncertainty in state transition rates.

4.1 Mathematical Preliminaries

A Markov Chain can be modeled by a stochastic process $X(t) : x \geq 0$, where $X(t)$ is a variable describing the occupied state over time with a random occurrence, assuming that its future evolution depends only on the present state, namely:

$$P(X_t = x_t | X_{t-1} = x_{t-1}, \dots, X_0 = x_0) = P(X_t = x_t | X_{t-1} = x_{t-1}) \tag{4.1}$$

Markov Chain time evolution can be modeled by both discrete or continuous stochastic processes, which can assume either a finite or an infinite number of states. The corresponding state transition probabilities are frequently assumed as time homogeneous. In particular, a discrete time Markov Chain (DTMC) is characterized by a set of N states, $S = \{s_1, \dots, s_N\}$, whose probabilities at a certain time step, t, are denoted as $\Pi(t) = \{\pi_1(t), \dots, \pi_N(t)\}$, while the corresponding state transition probabilities are denoted as p_{ij} for $i = 1, 2, \dots, N$ and $j = 1, 2, \dots, N$ (Luenberger, 1979). The state probabilities of a DTMC vary over time, and at each time step, the DTMC can only be in a single state.

The entire set of transition probabilities, which rules the DTMC dynamics, is described by a square matrix with elements belonging to $[0, 1]$, which is known as the *transition matrix*. Each row of this matrix $\mathbf{P}$, satisfies the following condition:

$$\sum_{j=1}^{n} p_{ij} = 1 \tag{4.2}$$

A DTMC is said to be *regular* if $\mathbf{P}^m > 0$ for each positive integer m. Regular DTMCs satisfy the *basic limit theorem for Markov Chain*, which makes it possible to obtain the steady-state probabilities by simply multiplying the transition matrix by itself enough times. Furthermore, for regular DTMCs, the steady-state probabilities are independent of their initial conditions, and the corresponding trajectories of each state probability can be computed by iteratively solving the following finite difference equation:

$$\pi(t + 1) = \mathbf{P} \cdot \pi(t) \tag{4.3}$$

A DTMC can be divided into communicating *classes*. A closed class is a set of communicating states, i.e. states that can switch from one to another in a finite number of steps. States inside a closed set cannot communicate to states outside the set itself. A closed class is also called *absorbing class*: once the process is in a state belonging to a closed class, it cannot leave the class anymore. If a closed class is made up by a single state, it is called *absorbing state*.

A continuous time Markov chain (CTMC) can be considered as a generalization of the DTMC in the continuous time domain. Indeed, while the state transitions in a DTMC can happen only at discrete time steps, in a CTMC they can happen at any time.

CTMC modeling requires properly redefining the transition matrix, by replacing the transition probabilities p_{ij} with transition rates q_{ij}. The latter are real-valued numbers that can vary in the interval $[0, \infty]$ $\forall i \neq j$ and $[-\infty, 0]$ $\forall i = j$. The diagonal elements q_{ii} of the transition rate matrix $\mathbf{Q}$ are equal to:

$$q_{ii} = -\sum_{k=1}^{n} q_{ik} \quad \forall k \neq i \tag{4.4}$$

where the sum of each row i should be equal to zero.

The trajectories of the state probabilities of a CTMC can be determined by solving the following set of ordinary differential equations:

$$\dot{\pi}(t) = \mathbf{Q}^T \cdot \pi(t) \tag{4.5}$$

where $\pi(t)$ and $\dot{\pi}(t)$ are the column vectors of state probabilities and their first derivatives at time t, respectively.

4.2 Effects of Data Uncertainties

The dynamics of Markov Chain-based models are mainly ruled by the transition matrix coefficients, which are conventionally modeled by real-valued deterministic parameters. This assumption cannot be verified in real application domains, where complex and correlated uncertainty sources could affect the determinism of the transition matrix. In this context, potential sources of uncertainties affecting the data-driven estimation of the transition rates derive from: (i) the scarcity of observed data; (ii) the complexities in measuring all the exogenous variables ruling the model dynamics; and (iii) the measurement noise and errors (Ethier and Kurtz, 2009).

Further sources of endogenous uncertainties are induced by the approximation errors of the model parameters identification algorithm and the quantization errors induced by discretizing continuous variables in a finite number of classes; these uncertainties can affect the level of accuracy of the estimated transition rates for both DTMCs and CTMCs. Moreover, as far as CTMCs are concerned, the data adopted for estimating the model parameters are generated by a sampling process, whose intrinsic errors induce other relevant uncertainties.

Finally, the input data adopted for model parameter estimation are frequently obtained by processing imprecise or qualitative information, which is difficult to model as probabilities.

Consequently, the development of nonprobabilistic frameworks for modeling and managing complex and correlated data uncertainties could be extremely useful for the dynamic analysis of uncertain Markov Chains.

In this context, a possible solution is to model the transition rates by interval variables and to apply IA-based operators to compute the outer estimation of the transient state probability trajectories. However, the large number of iterations required to compute the trajectory bounds could introduce excessive over-estimation errors (Chakraverty and Rout, 2020), therefore introducing spurious trajectories (Huang, 2018). Hence, the computed trajectory bounds could be much wider than the true solution bounds, especially in the presence of long computational chains in which the computed trajectory bounds could diverge (Brugnetti et al., 2020).

To mitigate these effects, AA-based computing is considered a promising research direction in Markov Chain-based models (Pepiciello et al., 2022).

4.3 Matrix Notation

In order to enhance the readability of the mathematical formulation, it is useful to represent vectors and matrices of affine form by affine vectors and matrices, respectively. In particular, an affine $(M \times R)$ matrix $\hat{A}$ with n uncertainty sources can be defined as:

$$\hat{A} = A_0 + A_1\varepsilon_1 + \cdots + A_n\varepsilon_n \tag{4.6}$$

where each element $\hat{a}_{ij}$ of the matrix $\hat{A}$ is:

$$\hat{a}_{ij} = a_{ij}^{(0)} + a_{ij}^{(1)}\varepsilon_1 + \cdots + a_{ij}^{(n)}\varepsilon_n \tag{4.7}$$

Hence, it follows that:

$$\hat{A} = \begin{bmatrix} a_{1,1}^{(0)} & a_{1,2}^{(0)} & \cdots & a_{1,R}^{(0)} \\ a_{2,1}^{(0)} & a_{2,2}^{(0)} & \cdots & a_{2,R}^{(0)} \\ \cdots & \cdots & & \\ a_{M,1}^{(0)} & a_{M,2}^{(0)} & \cdots & a_{M,R}^{(0)} \end{bmatrix} + \begin{bmatrix} a_{1,1}^{(1)} & a_{1,2}^{(1)} & \cdots & a_{1,R}^{(1)} \\ a_{2,1}^{(1)} & a_{2,2}^{(1)} & \cdots & a_{2,R}^{(1)} \\ \cdots & \cdots & & \\ a_{M,1}^{(1)} & a_{M,2}^{(1)} & \cdots & a_{M,R}^{(1)} \end{bmatrix} \varepsilon_1$$

$$+ \cdots + \begin{bmatrix} a_{1,1}^{(n)} & a_{1,2}^{(n)} & \cdots & a_{1,R}^{(n)} \\ a_{2,1}^{(n)} & a_{2,2}^{(n)} & \cdots & a_{2,R}^{(n)} \\ \cdots & \cdots & & \\ a_{M,1}^{(n)} & a_{M,2}^{(n)} & \cdots & a_{M,R}^{(n)} \end{bmatrix} \varepsilon_n$$

$$= A_0 + \sum_{k=1}^{n} A_k\varepsilon_k \tag{4.8}$$

Thus, if we define the vector $\varepsilon = \begin{bmatrix} 1 & \varepsilon_1 & \ldots & \varepsilon_n \end{bmatrix}$ and the vectorized deterministic matrix $\text{vec}(A) = \begin{bmatrix} A_0 & A_1 & \ldots & A_n \end{bmatrix}$, the affine matrix $\hat{A}$ can be expressed as a *Frobenius inner product* $\langle \text{vec}(A), \varepsilon \rangle_F$ (Jans, 1959). For example, given the affine matrix $\hat{B}$:

$$\hat{B} = \begin{bmatrix} 1 + 3\varepsilon_1 + 2\varepsilon_2 + 5\varepsilon_3 & 2 + 8\varepsilon_2 \\ 3 + 5\varepsilon_3 & 7\varepsilon_1 + 4\varepsilon_3 \end{bmatrix} \tag{4.9}$$

the coefficient matrices for each noise symbol would be:

$$\hat{B}_0 = \begin{bmatrix} 1 & 2 \\ 3 & 0 \end{bmatrix} \quad \hat{B}_1 = \begin{bmatrix} 3 & 0 \\ 0 & 7 \end{bmatrix}$$

$$\hat{B}_2 = \begin{bmatrix} 2 & 8 \\ 0 & 0 \end{bmatrix} \quad \hat{B}_3 = \begin{bmatrix} 5 & 0 \\ 5 & 4 \end{bmatrix} \tag{4.10}$$

The affine matrix $\hat{B}$ can also be expressed as $\langle \text{vec}(B), \varepsilon \rangle_F$.

4.4 AA-Based Uncertain Analysis

AA-based computing can be effectively applied to simulate Markov Chain-based models in the presence of uncertain transition rates/probabilities. For this, the real-valued transition matrix Q should be replaced by its affine counterpart $\hat{Q}$, whose components are no longer real numbers but affine forms modeling the uncertain model parameters. In particular, we will assume that each component of $\hat{Q}$ is affected by an independent source of uncertainty, which is modeled by a specific noise symbol. This assumption allows the affine matrix $\hat{Q}$ to be expressed as $\langle \text{vec}(\hat{Q}), \varepsilon \rangle_F$, where $\text{vec}(\hat{Q}) = \begin{bmatrix} Q_0 & Q_1 & \ldots & Q_n \end{bmatrix}$ is a tensor whose elements are the matrices Q_0 and Q_i $\forall\, i = 1, \ldots, n$, and $\varepsilon = \begin{bmatrix} 1 & \varepsilon_1 & \ldots & \varepsilon_n \end{bmatrix}$ is the vector representing the noise symbols.

Note that the assumption of considering an independent source of uncertainty for each element of the affine transition matrix does not affect the generality of the results. Indeed, modeling correlated uncertain transition rates is straightforward, as confirmed in the following example, where the uncertain coefficients of the affine transition matrix share the same noise symbol ε_1:

Example 4.1 Let's apply AA to model a two-state Markov Chain model of a repairable component in the presence of a single source of data uncertainty affecting both the repair rate (μ) and the fault rate (λ). The corresponding affine transition matrix can be defined as:

$$\hat{Q} = \begin{bmatrix} -\lambda_0 & \lambda_0 \\ \mu_0 & -\mu_0 \end{bmatrix} + \begin{bmatrix} -\lambda_1 & \lambda_1 \\ \mu_1 & -\mu_1 \end{bmatrix} \varepsilon_1 \tag{4.11}$$

where the affine forms describing the uncertain transition rates share the same noise symbol ε_1. In this case, the number of uncertain variables (i.e. λ and μ) is larger than the number of noise symbols, and the number of independent partial deviations is 2.

The definition of the affine transition matrix makes it possible to compute the affine forms describing the state probabilities at each time $t \neq t_0$, given their initial values π_0, by solving the following set of affine differential equations:

$$\dot{\hat{\pi}} = \hat{\pi} \cdot \hat{Q} \tag{4.12}$$

Note that the only nonaffine operator required to solve these differential equations is the AA-based multiplication, which introduces further noise symbols aimed at modeling the endogenous uncertainties generated during the computing process.

The solution of the set of affine differential equations formalized in (4.12) returns the central values and the partial deviations of the affine forms describing the trajectories of the state probabilities. To achieve this, the solution scheme proposed in Pepiciello et al. (2022) could be adopted for defining an equivalent deterministic system of ordinary differential equations of (4.12), which is obtained by deploying an integration scheme based on the explicit Euler algorithm and the AA-based multiplication operator described in Section 1.9.4.

In particular, the affine formula describing the explicit Euler integration is:

$$\Delta\hat{\pi} = \hat{\pi}(t) \cdot \hat{Q}\Delta t \tag{4.13}$$

where the affine form affine forms $\Delta\hat{\pi}$ and $\hat{\pi}$ can be described as follow:

$$\Delta\hat{\pi} = \pi_0(t+1) - \pi_0(t) + (\pi_1(t+1) - \pi_1(t))\varepsilon_1 + \cdots + (\pi_{n+1}(t+1)$$
$$- \pi_{n+1}(t))\varepsilon_{n+1}$$
$$\hat{\pi} = \pi_0(t) + \pi_1(t)\varepsilon_1 + \cdots + \pi_{n+1}(t)\varepsilon_{n+1} \tag{4.14}$$

where $\varepsilon_1 \ldots \varepsilon_n$ denote the primitive noise symbols, i.e. those modeling the input data uncertainties, and ε_{n+1} is the noise symbol modeling the cumulative effects of the endogenous uncertainties, which are induced by the approximation errors of all the AA-based multiplications.

Moreover, it is useful to define the affine matrix $\hat{\pi}$, whose rows are the partial deviations of the state probabilities for each noise symbol.

Hence, Eq. (4.13) can be expressed by a Frobenius inner product as:

$$\Delta\hat{\pi} = \langle\langle\mathbf{\Pi}(t), \varepsilon\rangle_F, \langle\text{vec}(Q), \varepsilon\rangle_F\rangle_F \Delta t \tag{4.15}$$

or equivalently:

$$\Delta\hat{\pi} = (\pi_0(t) + \pi_1(t)\varepsilon_1 + \cdots + \pi_{n+1}(t)\varepsilon_{n+1}) \cdot (Q_0 + Q_1\varepsilon_1 + \cdots + Q_n\varepsilon_n)\Delta t \tag{4.16}$$

which allows $\Delta\hat{\pi}$ to be expressed as follows:

$$\Delta\hat{\pi} = \pi_0 \cdot Q_0 + \sum_{i=1}^{n}(\pi_0 \cdot Q_i + \pi_i \cdot Q_0)\varepsilon_i + \sum_{i=1}^{n}\pi_i \cdot Q_i\varepsilon_i^2$$

$$+ \sum_{i=1}^{n}\sum_{j=1, j\neq i}^{n}(\pi_i \cdot Q_j + \pi_j \cdot Q_i)\varepsilon_i\varepsilon_j$$

$$+ \left(\pi_{n+1} \cdot Q_0 + \sum_{i=1}^{n}Q_i \cdot \pi_{n+1}\varepsilon_i\right)\varepsilon_{n+1} \qquad (4.17)$$

Then, by combining (4.14) and (4.17), it is possible to formalize the following set of deterministic equations describing the time evolution of the central values and partial deviations of the primitive noise symbols of the affine state probabilities:

$$\begin{cases} \pi_0(t+1) = \pi_0(t) \cdot Q_0\Delta t + \pi_0(t) \\ \pi_i(t+1) = (\pi_0(t) \cdot Q_i + \pi_i(t) \cdot Q_0)\Delta t + \pi_i(t) \quad \forall i \in [1, n] \end{cases} \qquad (4.18)$$

Furthermore, in order to model the effects of higher-order combinations of the noise symbols in (4.17), which are described by the following quadratic form:

$$f(\varepsilon) = \sum_{i=1}^{n}\pi_i \cdot Q_i\varepsilon_i^2 + \sum_{i=1}^{n}\sum_{\substack{j=1 \\ j\neq i}}^{n}(\pi_i \cdot Q_j + \pi_j \cdot Q_i)\varepsilon_i\varepsilon_j + \sum_{i=1}^{n}\pi_{n+1} \cdot Q_i\varepsilon_i\varepsilon_{n+1}$$

$$(4.19)$$

The following deterministic equations should be integrated into (4.17):

$$\pi_{n+1}(t+1) = (\pi_{n+1}(t) \cdot Q_0 + \Delta\pi_{nl})\Delta t + \pi_{n+1}(t) \qquad (4.20)$$

where $\Delta\pi_{nl}$ can be obtained by computing an outer bound estimation of the non-affine function $f(\varepsilon)$ (e.g. by using the Chebyshev approximation), namely:

$$\Delta\pi_{nl} = \frac{\max(f(\varepsilon)) - \min(f(\varepsilon))}{2} \qquad (4.21)$$

Consequently, the time-evolution of the affine vectors $\hat{\pi}(t)$ can be computed as:

$$\hat{\pi}(t) = \pi_0(t) + \sum_{i=1}^{n+1}\pi_i(t)\varepsilon_i \quad \forall t \geq 0 \qquad (4.22)$$

and the corresponding bounds are :

$$\bar{\pi}(t) = \left[\pi_0(t) - \sum_{i=1}^{n+1}|\pi_i(t)|, \pi_0(t) + \sum_{i=1}^{n+1}|\pi_i(t)|) = \right] \qquad (4.23)$$

Finally, it is worth noting that the first set of equations formalized in (4.18) describes the state probability trajectories of the Markov Chain model when all the uncertain variables are at their nominal (central) values, while the other set of $n + 1$ equations describes the effects of both the endogenous and exogenous uncertainties on the nominal trajectories.

4.5 Application Examples

In order to prove the effectiveness of the described AA-based computing method to simulate Markov Chain-based models in the presence of transition rate uncertainties, the results of two application examples are presented and discussed. This first example considers the use of CTMCs for grid resilience analysis (Pepiciello et al., 2022), while the second utilizes CTMCs to model energy storage (Song et al., 2012). In these studies, the CTMCs are simulated by assuming a fixed integration step of 0.001 unit of time, and the solution bounds obtained using AA are benchmarked with those computed by a Monte Carlo-based method. The Monte Carlo method implemented samples of the input data uncertainty and, for each sample, generates the corresponding transition matrix and computes the state probability trajectories. As far as the input data uncertainties are concerned, transition rate tolerances of $\pm10\%$ and $\pm20\%$ have been considered.

4.5.1 Case Study 1: Grid Resilience Analysis

This example deals with the application of CTMCs for modeling the operation of a network of grid-connected microgrids in the presence of contingencies (Liu et al., 2016). The operation states of this system can be classified as:

- s_1 (*normal state*): The power system is under normal operating conditions, and all microgrids are grid-connected.
- s_2 (*islanded state*): In order to react to severe grid faults, the microgrids are disconnected from the main grid, hence operating in islanded mode, i.e. used to supply the local power demand and controlling the frequency and voltage within the microgrid.
- s_3 (*contingency state*): The power system is congested due to the inability of some microgrids to supply the local power demands, hence inducing grid overloads.

Thus, in the presence of a severe grid fault, the system will evolve from state s_1 to s_2 and, consequently, to state s_3 or s_1, dependent on if the time required to repair

Table 4.1 Transition rate matrix.

From/to	s_1	s_2	s_3
s_1	−1.00	1.00	0.00
s_2	26.10	−78.20	52.10
s_3	17.38	17.38	−34.76

the fault is longer/shorter than the microgrids runtime, respectively. The corresponding nominal (central) values of the transition rates are reported in Table 4.1 (Pepiciello et al., 2022).

To model the uncertainty affecting these parameters, six noise symbols, i.e. one for each independent element of the transition rate matrix, have been considered. The corresponding partial deviations can be determined on the basis of the fixed tolerance, which has been assumed ±20%. The system is assumed to start in state s_2 for $t = 0$, hence $\pi(0) = [0\ 1\ 0]$.

The corresponding time-evolution of the central values and bounds of the state probabilities computed by AA and the Monte Carlo-based methods are reported in Figure 4.1.

When analyzing these results it is worth noting that, as expected, the solution bounds computed by applying AA are outer estimations of the trajectories computed by the Monte Carlo model, hence resulting in robust estimations of the transient state probabilities. This important feature, which derives from the Fundamental Theorem of Range Analysis, is extremely useful in supporting decision making under risk. This is confirmed in the second application example.

4.5.2 Case Study 2: Energy Storage Model

This example aims at modeling the State of Charge (SoC) of an energy storage system connected to a renewable power generator by CTMC (Song et al., 2012). In this example, the allowable SoC range has been discretized into five classes, each one associated to a CTMC state. The corresponding transition matrix is reported in Table 4.2 (Pepiciello et al., 2022):

The CTMC dynamics depends on the SoC evolution, which is influenced by the uncertain generation/demand profiles. By representing the transition rates by affine forms, it is possible to model complex phenomena such as the randomness of the renewable power-generated profiles and the impacts of component performance degradation during their life cycle (e.g. storage capacity, generator efficiency) on the charging/discharging profiles. In this example, these phenomena

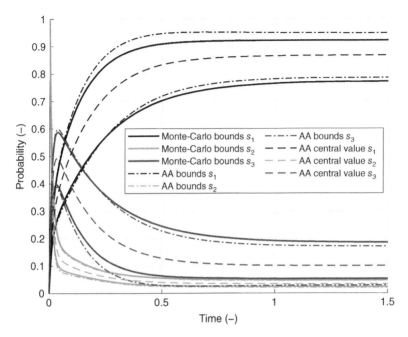

Figure 4.1 Case Study 1: Transient state probabilities bounds computed by AA and Monte Carlo for an input data tolerance of $= \pm 20\%$.

Table 4.2 Transition rate matrix.

From/to	s_1	s_2	s_3	s_4	s_5
s_1	−2.40	1.60	0.80	0.00	0.00
s_2	1.60	−4.00	1.60	0.80	0.00
s_3	0.40	1.20	−4.00	1.60	0.80
s_4	0.00	0.40	1.20	−4.00	2.40
s_5	0.00	0.00	0.40	1.20	−1.60

have been modeled by assuming a tolerance of $\pm 20\%$ for each independent component of the transition matrix.

The transient evolution of the state probabilities obtained by applying the described AA-based technique and the Monte Carlo-based method, assuming s_1 as the initial CTMC state, is reported in Figure 4.2. The analysis of this figure confirms that, also in this case, the trajectories computed by applying AA-based computing are outer estimations of the corresponding solution bounds obtained by the sampling-based approach.

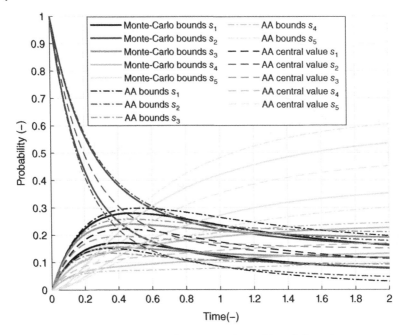

Figure 4.2 Case Study 2: Monte-Carlo and Affine Arithmetic bounds, uncertainty = ±20%.

4.5.3 Summary

Finally, as far as the conservativism level of AA is concerned, it is worth noting that:

- The solution domains computed by AA are polyhedrons, which are guaranteed to include the real solution sets. Consequently, AA-based computing tends to overestimate the real solution bounds.
- AA-based solutions consider the effects of the worst-case combination of the input data uncertainties, whose occurrence probability could be extremely low and, hence, hardly identified by sample-based approaches.
- The deployment of nonaffine operators in iterative computing chains introduce approximation errors, whose cumulative effects increase the endogenous uncertainties.

5

Small-Signal Stability Analysis of Uncertain Power Systems

Small-signal stability analysis of power system aims at assessing the ability of the grid to maintain synchronism in the presence of small disturbances, such as those induced by moderate variations of the power demand/generation profiles. This is a fundamental prerequisite for reliable power system operation, as systems that become unstable in terms of small-signal stability cannot operate under the corresponding operating conditions.

This analysis is becoming increasingly important in modern power systems, where the decommissioning and replacement of large synchronous generators with power electronics-interfaced renewable power generators is making power systems more vulnerable to both internal and external dynamic perturbations.

In this context, the growing level of model and data uncertainties that affect real-time power system operation is another relevant issue to address for enhancing grid resilience against dynamic perturbations. In particular, data uncertainties stem from several sources and influence both the structure and parameters of the equations characterizing the loads and distributed generators dynamics (Hiskens and Alseddiqui, 2006). They may include both aleatory (e.g. irreducible) uncertainties, which are intrinsically related to the inherent randomness and natural stochasticity characterizing the analyzed power system, and epistemic (e.g. reducible) uncertainties, which are mainly related to measurement errors and unreliable model parameter estimations (Preece and Milanović, 2015).

These uncertainties render it difficult, or even impossible, to reliably assess grid robustness against dynamic perturbations using deterministic small-signal stability analysis. If not correctly considered, there is a risk of exposing potential grid vulnerabilities to multiple and correlated uncertain events (Rueda et al., 2009).

For this purpose, conventional solution techniques try to address small-signal stability analysis for the worst-case combination of the input data uncertainties, without modeling their stochastic nature (Rueda et al., 2009).

More sophisticated solution approaches try to compute the sensitivity of the system matrix eigenvalues to system parameters using probabilistic techniques (Burchett and Heydt, 1978); these can be classified into numerical and analytical methods (Pareek and Nguyen, 2020). The first class of methods aims at defining and analyzing a proper subset of all scenarios in order to infer the statistical features of the system matrix eigenvalues. In this context, sampling-based approaches, such as those based on Monte Carlo simulations, are among the most used solution approaches (Xu et al., 2006, 2005). Unfortunately, the computational burden required to obtain reliable solutions using these methods could be prohibitively high, particularly for large-scale systems. Hence, alternative techniques for properly reducing the sample space have been conceptualized to consider the trade-off between algorithm accuracy and complexity (Huang et al., 2013).

As far as analytical methods are concerned, they try to formalize and solve simplified analytical expressions which directly correlate the input data uncertainty with the system matrix eigenvalues. For this purpose, the point estimate method is frequently applied, since it can significantly reduce the computational burden (requiring only $2n + 1$ deterministic analyses for characterizing the statistical features of n uncertain variables) (Preece et al., 2014).

Other analytic techniques often used for uncertain small-signal stability analysis are the cumulant method, which aims at computing the cumulants of the probability distributions of the system matrix eigenvalues, rather than their moments (Bu et al., 2014), and the collocation method, which allows approximating the system matrix eigenvalues by polynomial expansion (Preece et al., 2012).

However, reliable estimations of the parameters of the probability distributions and the difficulty in modeling the statistical dependence between the input data uncertainties are two of the main limitations of these techniques.

To overcome these issues, nonprobabilistic-based techniques have been adopted for modeling and managing data uncertainties in small-signal stability analysis. In particular, in Pan et al. (2017), a robust stable regionwise approach was proposed for estimating the stable system domain, while in Ahmadi and Kazemi (2020) a fuzzy sets-based possibilistic method was developed for assessing the small-signal stability of microgrids.

In this context, affine arithmetic (IA) has been recognized as an enabling methodology for reliably estimating the bounds of the system matrix eigenvalues in the presence of data uncertainty (Pepiciello and Vaccaro, 2022). To achieve this, in section 5.1 small-signal stability analysis is formalized as an interval eigenvalue problem, which can be solved by the IA-based processing tools described in section 5.2.

5.1 Problem Formulation

Power system stability is defined in Kundur et al. (2004) as "the ability of the system, for a given initial operating condition, to return to a state of operating equilibrium after a physical disturbance, with most of the variables of the system constrained so that the entire system remains intact," and can be classified by frequency, voltage, resonance, and rotor angle stability (Giannuzzi et al., 2022).

In this context, small-signal stability is a fundamental preliminary analysis, which aims at assessing the grid stability in the presence of small dynamic perturbations (Milano et al., 2018). For this purpose, the power system dynamic behavior is modeled by the following set of nonlinear differential-algebraic equations (DAE), whose algebraic and dynamic variables describe the quasi-steady system operation and the corresponding transient evolution, respectively Sauer et al. (2017).

$$\dot{x} = f(x, y)$$
$$0 = g(x, y) \tag{5.1}$$

where y and x are the dynamic and algebraic variable vectors, while $g(x, y)$ and $f(x, y)$ are sets of algebraic and nonlinear differential equations, respectively.

Small-signal stability analysis requires linearizing the set of equations around an equilibrium point (x_0, y_0), hence obtaining:

$$\Delta \dot{x} = A \Delta x + B \Delta y$$
$$0 = C \Delta x + D \Delta y \tag{5.2}$$

where the matrices A, B, C, and D can be determined as follows:

$$A = \frac{\partial f}{\partial x}(x_0, y_0) \qquad B = \frac{\partial f}{\partial y}(x_0, y_0)$$
$$C = \frac{\partial g}{\partial x}(x_0, y_0) \qquad D = \frac{\partial g}{\partial y}(x_0, y_0) \tag{5.3}$$

Observe that the definition of the following matrix:

$$A_{sys} = A - BD^{-1}C \tag{5.4}$$

allows the problem to be simplified by eliminating the algebraic variables, and the stability assessment of (5.2) to be performed by solving the following eigenvalue problem:

$$A_{sys} u = \lambda u \tag{5.5}$$

where u is the eigenvector associated to the eigenvalue λ.

Indeed, the system described by (5.2) is stable if and only if the real part of all the eigenvalues is negative. Moreover, eigenvalues analysis makes it possible to infer other useful information about the critical events that could threaten power system operation. In particular, even for stable power system, the presence of some eigenvalues closer to the instability region could be considered as a reliable indicator of potential system vulnerabilities. For this purpose, analyzing the damping ratio ζ_i of each complex eigenvalue $\lambda_i = \alpha_i + j\beta_i$ has been recognized as a valuable tool to enhance situational awareness.

$$\zeta_i = \frac{-\alpha_i}{\sqrt{\alpha_i^2 + \beta_i^2}} \tag{5.6}$$

In particular, complex eigenvalues characterized by $\zeta_i \leq 5\%$ are considered as a reliable indication of the presence of potential critical modes (Rogers, 2012).

5.2 The Interval Eigenvalue Problem

Probabilistic-based models are not the only way to handle uncertainty; uncertain variables can also be represented by defining proper domains containing all the possible values that these variables can assume. When defining these domains, some hypotheses about the uncertain variables are frequently assumed (Ben-Haim and Elishakoff, 2013). In particular, each uncertain variable x is bounded, namely $|x| < a$, where a is a finite positive number, its range is defined by two deterministic functions, namely $x_{lower}(t) \leq x(t) \leq x_{upper}(t)$, and it has an integral square bound, namely $\int_{-\infty}^{\infty} x^2(t)dt \leq a$.

By making these assumptions, it is possible to model uncertain variables using a set-theoretic approach. This method removes the need to make assumptions about the probability distributions of the uncertain variables, making it a viable alternative to probabilistic techniques, particularly when only limited information about the uncertain variables is available.

This approach to the interval eigenvalue problem can be formalized by the following set of linear interval equations: Deif (1991):

$$A^I x = b^I \tag{5.7}$$

where the interval matrix A^I and the interval vector b^I can be expressed as

$$A^I = [A^C - \Delta A, A^C + \Delta A] \tag{5.8}$$

$$b^I = [b^C - \Delta b, b^C + \Delta b] \tag{5.9}$$

where, as usual, A^C and b^C are the central values of A^I and b^I, respectively, while ΔA and Δb are their corresponding radius.

Solving the problem (5.7) requires the computation of the following solutions set (Oettli and Prager, 1964):

$$X = \{x : Ax = b, A \in A^I, b \in b^I\} \tag{5.10}$$

Note that $x \in X$ only when the following equation is satisfied (Oettli and Prager, 1964):

$$|A^C x - b^C| \le \Delta A|x| + \Delta b \tag{5.11}$$

Therefore, an important problem to solve in the context of uncertain small-signal stability analysis is to find the set of interval eigenvalues Γ of the interval matrix A^I, namely:

$$\Gamma = \{\lambda \in \mathbb{C}, Ax = \lambda x, x \ne 0, A \in A^I\} \tag{5.12}$$

Solving this problem, which bounds the eigenvalues domain of the uncertain matrix A, can be achieved using the results of the following theorems (Muscolino and Sofi, 2012; Qiu et al., 2005):

Theorem 5.1 *If A^I is a real interval matrix and s_{kj}^i, evaluated at A^C, is constant over A^I, then the real part of eigenvalue λ_r^i of $A \in A^I$ ranges over the interval:*

$$\lambda_r^{i,I} = [\lambda_r^i(A^C - \Delta A \cdot S^i), \lambda_r^i(A^C + \Delta A \cdot S^i)] \tag{5.13}$$

where S^i is a scalar matrix containing the elements: $s_{kj}^i = sgn(y_{rk}^i x_{rj}^i + y_{yk}^i x_{yj}^i)$ and $x^i = x_r^i + jx_y^i$, $y^i = y_r^i + jy_y^i$ are the normalized right and left eigenvectors of A^C associated with the ith eigenvalue.

Theorem 5.2 *If A^I is a real interval matrix and s_{kj}^i, evaluated at A^C, is constant over A^I, then the imaginary part of eigenvalue λ_y^i of $A \in A^I$ ranges over the interval:*

$$\lambda_y^{i,I} = [\lambda_y^i(A^C - \Delta A \cdot S^i), \lambda_y^i(A^C + \Delta A \cdot S^i)] \tag{5.14}$$

where S^i is a scalar matrix containing the elements: $s_{kj}^i = sgn(y_{rk}^i x_{yj}^i + y_{yk}^i x_{rj}^i)$ and $x^i = x_r^i + jx_y^i$, $y^i = y_r^i + jy_y^i$ are the normalized right and left eigenvectors of A^C associated with the ith eigenvalue.

These theorems make it possible to calculate outer estimates of the eigenvalue bounds by solving standard deterministic problems. As a result, it is possible to determine the small signal stability of the power system for all possible combinations of input data uncertainty.

5.3 Applications

5.3.1 Case Study 1

The first application example deals with the small-signal stability analysis of the single machine infinite bus model described in Kundur et al. (1994), whose system matrix is detailed in (5.15).

$$
\begin{bmatrix} \Delta\dot{\omega} \\ \Delta\dot{\delta} \\ \Delta\dot{\psi} \end{bmatrix} = \begin{bmatrix} 0 & -0.109 & -0.123 \\ 376.991 & 0 & 0 \\ 0 & -0.194 & -0.423 \end{bmatrix} \begin{bmatrix} \Delta\omega \\ \Delta\delta \\ \Delta\psi \end{bmatrix} \tag{5.15}
$$

In this equation $\Delta\omega$ and $\Delta\delta$ are the rotor speed and angle variation, respectively, while $\Delta\psi$ is the flux linkage. The parameters of this dynamic model are influenced by the dynamic properties of the synchronous generator, such as its inertia, damping, reactance, and excitation system parameters.

In the deterministic case, the eigenvalues of the system matrix are $\lambda_{1,2} = -0.109 \pm j6.568$, $\lambda_3 = -0.202$ (Pepiciello and Vaccaro, 2022).

Let us now consider the effects of data uncertainty by assuming a $\pm 10\%$ tolerance on the inertia constant, which affects the determinism of the components of the first row of the dynamic matrix, whose radius ΔA is

$$
\Delta A = \begin{bmatrix} 0 & 0.0115 & 0.0130 \\ 0 & 0 & 0 \\ 0 & 0 & 0 \end{bmatrix} \tag{5.16}
$$

The application of the described method requires determining the matrix S^i, for the ith eigenvalue of A, and the corresponding left x^i and right y^i eigenvectors. Then, the minimum and maximum bounds of the interval eigenvalues can be computed using (5.13). The results obtained from this method and those from a Monte Carlo-based method are presented in Table 5.1.

When examining these results, it is worth noting that the IA-based method can accurately enclose the solution bounds for all possible values of the inertia constant.

Once the bounds of the eigenvalues have been obtained it is possible to calculate the corresponding bounds of the damping ratio for the oscillating mode. This can be computed by performing (5.6) using IA-based operators, resulting in $\zeta^I = [\zeta_{min}, \zeta_{max}] = [0.0113, 0.0236]$. From this result, it can be inferred that for certain values of generator inertia, the oscillating mode could be critical, hence, additional damping is necessary.

The computed solution bounds can also provide insight into the sensitivity of the eigenvalues' real and imaginary parts to input data uncertainty, as the area of the rectangle representing the interval eigenvalues is a reliable indicator.

Table 5.1 Eigenvalues bounds.

Deterministic	IA-based method	Monte Carlo
$-0.109 + j6.568$	$[-0.136, -0.083] \pm j[5.769, 7.351]$	$[-0.133, -0.090] \pm j[6.234, 6.861]$
$-0.109 - j6.568$	$[-0.136, -0.083] \pm j[-7.351, -5.769]$	$[-0.133, -0.090] \pm j[-6.861, -6.234]$
-0.202	$[-0.255, -0.149]$	$[-0.240, -0.154]$

5.3.2 Case Study 2

The second application example is based on the small-signal stability analysis of the New England 39 bus system. This system consists of 10 generators with automatic voltage regulation and power system stabilizers, interconnected by 46 power lines and transformers. The characteristic data of the test system are reported in Canizares et al. (2015).

The initial system operation state can be computed using a power flow analysis tool (Zimmerman et al., 2010). The system matrix A_{sys}, which describes the dynamic evolution around this operating point, can then be obtained as described in Sauer et al. (2017).

A realistic assumption of $\pm 5\%$ tolerance for the inertia constants of each generator has been made in regards to data uncertainty, taking into account the effects of renewable power generators on power system inertia (Shim et al., 2020).

By using the matrix ΔA_{sys} that accounts for uncertainty and the matrices S_i, one for each eigenvalue λ_i as in equations (5.13) and (5.14), the interval eigenvalues λ_i^I have been calculated.

Table 5.2 Eigenvalues bounds.

Modes	Interval eigenvalues	Damping ratio
1	$[-0.321, -0.276] \pm j[10.049, 10.547]$	$[0.026, 0.031]$
2	$[-0.444, -0.102] \pm j[9.369, 9.923]$	$[0.010, 0.047]$
3	$[-0.423, -0.088] \pm j[9.389, 9.888]$	$[0.009, 0.045]$
4	$[-0.295, -0.134] + j[8.872, 9.403]$	$[0.015, 0.033]$
5	$[-0.294, -0.201] + j[8.873, 9.400]$	$[0.021, 0.033]$
6	$[-0.307, -0.083] + j[7.481, 8.478]$	$[0.010, 0.041]$
7	$[-0.265, -0.110] + j[7.139, 8.322]$	$[0.013, 0.037]$
8	$[-0.324, -0.122] + j[7.106, 8.087]$	$[0.015, 0.045]$
9	$[-0.142, -0.041] + j[3.095, 3.529]$	$[0.011, 0.046]$
10	$[-0.169, -0.096] + j[2.423, 2.656]$	$[0.036, 0.068]$

Table 5.2 summarizes the obtained results, which include the bounds of the first critical interval eigenvalues and their interval damping ratios. The obtained results show that the uncertainty sources that impact the inertia constant have an effect on both the real and imaginary components of the eigenvalues, and the sensitivity of this effect varies depending on the specific eigenvalue being considered.

Furthermore, the study of the bounds of the damping ratios makes it possible to accurately identify the most critical modes and evaluates the significant effects caused by the uncertainty in the inertia constant, which can result in changes from nonoscillatory to oscillatory modes.

Lastly, as also outlined in Pepiciello and Vaccaro (2022), the computed solution bounds are a reliable outer estimation of the solution bounds obtained using a Monte Carlo model, which requires larger computational burden compared to IA. Indeed, the described approach only requires computing the eigenvalues of eight deterministic matrices, according to Eqs. (5.13) and (5.14). This is independent of the number of uncertain variables and the percentage of uncertainty affecting each of them.

6

Uncertain Power Components Thermal Analysis

The protection of power system components is crucial in the open electricity market, where any power system failure can cause severe harm to the many operators who share the power system infrastructure. In the past, to ensure dependable and safe service, asset owners have used conservative thermal ratings, calculated assuming over-conservative weather conditions (such as low wind speeds and high ambient temperatures), to ascertain loading limits of power system components.

The deployment of this worst-case strategy reduces the chances of system failure, but it also leads to underutilization of transmission and distribution assets (Study Committee 23-CIGRE, 202). In the current competitive environment, where power companies are encouraged to fully use their equipment to increase transmission capacity and maximize revenue through long-distance energy exchange, this conservative approach is no longer adequate (Piccolo et al., 2004).

To increase grid flexibility, it is necessary to have a more accurate evaluation of the loading of power system components and a better management of associated risks. This requires accurate computation of actual thermal ratings and predictions of how they will change over time, at timescales spanning from a few minutes up to several hours (Villacci and Vaccaro, 2007).

Due to their fundamental role in the transmission and distribution of power, there is particular interest in the thermal analysis of overhead lines and cables, and an accurate prediction of the conductor temperature evolution is essential information for effectively evaluating the risk related to certain loading policies. Specifically, this requires the utilization of a suitable thermal model capable of predicting the evolution of the conductor temperature for each hypothetical loading level and the corresponding maximum time duration, as a function of the actual conductor thermal state and the forecast environmental conditions.

Interval Methods for Uncertain Power System Analysis, First Edition. Alfredo Vaccaro.
© 2023 The Institute of Electrical and Electronics Engineers, Inc. Published 2023 by John Wiley & Sons, Inc.

In this context, various mathematical models have been developed, based both on simplified heat transfer equations and detailed thermal modeling (E8, 1993). These models calculate the conductor temperature profile by solving the balance equations that govern heat exchange between the conductor and the environment.

Both simplified and detailed thermal models require some specific input data, which can be affected by many uncertainties. Accordingly, they could exhibit strong sensitivity to parameter variations. Large uncertainties arise from several sources, such as the level of atmospheric pollution, aging, spatial distribution of climatic variables. In view of these uncertainties, the estimated thermal rating of the power system components can be very different from the reality. Therefore, in order to achieve acceptable accuracy and robustness in thermal rating prediction, it is necessary to take these sources of uncertainty into account. To this end, many detailed probabilistic methods, accounting for the variability and stochastic nature of the model input parameters, have been proposed in several papers (Black et al., 1988).

While these methodologies are useful, particularly for planning studies, they have some limitations, primarily due to the nonnormal distribution and statistical correlations of weather variables and the difficulty of modeling some input data using probability distributions. These limitations can lead to time-consuming calculations, particularly when it comes to the thermal rating assessment of large-scale power networks.

To address some of these limitations, self-validating computation can be employed. The main advantage of these techniques is that the solution algorithm automatically keeps track of the accuracy of the computed quantities as part of the calculation process without the need for information about the uncertainty in the parameters.

6.1 Thermal Rating Assessment of Overhead Lines

The steady-state temperature of an overhead conductor can be modeled under fixed values of wind velocity and direction, ambient temperature, solar radiation, and load current using the following equation:

$$q_c(T_c, T_a, V_w, \theta) + q_r(T_c, T_a) = q_s + I^2 R(T_c) \tag{6.1}$$

where

- T_c and T_a are the conductor and ambient temperature;
- V_w and θ are the wind speed and direction;
- q_c is the convection heat flux;
- q_r is the radiated heat flux;

- q_s is the solar heat gain;
- I and R are the conductor current and resistance.

Under these operating conditions, the maximum current that the line can carry continuously without exceeding its maximum allowable conductor temperature $T_{c,max}$ can be determined by inverting (6.1):

$$I_{max} = \sqrt{\frac{q_c(T_c, T_a, V_w, \theta) + q_r(T_c, T_a) - q_s}{R(T_c)}} \tag{6.2}$$

This value is often referred to as the ampacity or the static thermal rating of the overhead line.

When the line current or the environmental conditions change, the conductor thermal dynamics are perturbed, and the resulting changes in temperature can be described using the following differential equation:

$$\frac{dT_c}{dt} = \frac{1}{mC_p} \left[q_s + I^2 R(T_c) - q_c(T_c, T_a, V_w, \theta) - q_r(T_c, T_a) \right] \tag{6.3}$$

In this operating condition, transient thermal ratings can be determined by solving this differential equation for various load current values, and then identifying the current value which causes the conductor temperature to reach its maximum allowable threshold within the given time frame.

The assessment of steady state and transient thermal ratings requires all the phenomena ruling the conductor heat exchange to be characterized.

In particular, the convective term q_c can be calculated based on the wind speed by utilizing one of the following equations, which are valid for zero $q_{c,0}$, low $q_{c,low}$, and high $q_{c,high}$ wind speeds, respectively E8 (1993):

$$q_{c,0} = 0.283 \rho_f^{0.5} D^{0.75} (T_c - T_a)^{1.25}$$

$$q_{c,low} = K_{angle} \left[1.01 + 0.371 \left(\frac{D\rho_f V_w}{\mu_f} \right)^{0.25} \right] k_f (T_c - T_a) \tag{6.4}$$

$$q_{c,high} = K_{angle} 0.1695 \left(\frac{D\rho_f V_w}{\mu_f} \right)^{0.6} k_f (T_c - T_a)$$

where

- D is the conductor diameter;
- ρ_f, μ_f, and k_f are the density, viscosity, and coefficient of thermal conductivity of air, respectively, which can be determined by using the following regressive functions:

$$\rho_f = \frac{0.080695 - 0.2901 \times 10^{-5} H_c + 0.37 \times 10^{-5} H_c^2}{1 + 0.00367 T_{film}}$$

$$\mu_f = 0.0415 + 1.2034 \times 10^{-4} \, T_{film} - 1.1442 \times 10^{-7} \, T_{film}^2 + 1.9416$$
$$\times 10^{-10} \, T_{film}^3 \tag{6.5}$$

$$k_f = 0.007388 + 2.27889 \times 10^{-5} \, T_{film} - 1.34328 \times 10^{-9} \, T_{film}^2$$

$$T_{film} = \frac{T_c + T_a}{2}$$

where H_c is the sun altitude;

- K_{angle} is the wind direction factor, which can be determined as

$$K_{angle} = 1.194 - \cos(\phi) + 0.194 \cos(2\phi) + 0.368 \sin(3\phi) \tag{6.6}$$

where ϕ is the angle between the conductor axis and the wind direction.

The radiated heat loss can be computed by using the following equation:

$$q_r = 0.138 D \tau \left[\left(\frac{T_c + 273}{100} \right)^4 - \left(\frac{T_a + 273}{100} \right)^4 \right] \tag{6.7}$$

where τ is the coefficient of emissivity.

The solar heat gain can be determined as follows:

$$q_s = \alpha Q_s \sin(\Theta) A$$
$$\Theta = \cos^{-1}(\cos(H_c) \cos(Z_c - Z_L)) \tag{6.8}$$

where A is the projected area of the conductor, Z_c and Z_L are the sun and line azimuth, respectively, α is the solar absorptivity, and Q_s is the total heat flux, which can be estimated by using regressive equations of the solar altitude (E8, 1993).

6.1.1 Sources of Data Uncertainties

Static and transient thermal rating assessment of overhead lines requires the solution of conductor thermal models, whose parameters can be affected by strong and correlated uncertainties. These uncertainties mainly derive from the complexities when modeling the effects of conductor aging, atmospheric pollution, and wind direction on the radiated heat flux and solar heat gain (Piccolo et al., 2004).

In particular, the radiated heat loss depends on the actual state of the conductor surface via the coefficient of emissivity τ, which can vary in the interval $[0.23, 0.91]$. Moreover, the solar heat gain is influenced by the conductor surface aging and the atmospheric pollution via the absorption coefficient α, which can vary in the interval $[0.23, 0.97]$.

Other relevant sources of uncertainty are induced by the values of the weather variables used for computing the thermal ratings (e.g. wind speed and direction). Indeed, although these variables can be measured at specific locations, they exhibit considerable spatial variations (Reding, 1994). Hence, local measurements may not be representative of the actual heat exchange conditions along the entire line route. Consequently, the thermal ratings can vary along the line route, and the line loading is limited by the rating computed for the critical span (i.e. the span characterized by the worst heat exchange conditions) (Douglass, 1988).

Moreover, transmission system operators are interested in determining both the actual and the predicted thermal ratings over time horizons varying from one to six hours ahead. Hence, the weather variables used in thermal rating prediction are estimated using forecasting algorithms, whose prediction errors introduce additional exogenous uncertainty sources.

All these uncertainties affect the determinism of the input data, and, consequently, the tolerance of the computed thermal rating.

6.1.2 AA-Based Thermal Rating Assessment

Affine arithmetic (AA) can be applied for reliable computing of both static and dynamic thermal ratings in the presence of data uncertainties. For this purpose, all of the uncertain input parameters should be represented by affine forms, whose noise symbols model the following independent sources of data uncertainty (Piccolo et al., 2004):

- conductor aging, ε_{age}
- atmospheric pollution, ε_{pol}
- spatial distribution of the weather variables along the line route, ε_{spt}
- forecasting error of the ambient temperature, ε_{frc1}
- forecasting error of the wind speed, ε_{frc2}
- forecasting error of the wind direction, ε_{frc3}
- forecasting error of the line current, ε_{frc4}

Note that four different noise symbols have been considered for modeling the uncertainties induced by the forecasting errors of the ambient temperature, wind speed, wind direction, and line current, which are predicted by using different forecasting algorithms.

Hence, the affine forms modeling the uncertain input data are

$$\hat{\alpha} = \alpha_0 + \alpha_{age}\varepsilon_{age} + \alpha_{pol}\varepsilon_{pol}$$

$$\hat{\tau} = \tau_0 + \tau_{age}\varepsilon_{age} + \tau_{pol}\varepsilon_{pol}$$

$$\hat{Q}_s = Q_{s,0} + Q_{s,pol}\varepsilon_{pol}$$

$$\hat{T}_a = T_{a,0} + T_{a,spt}\varepsilon_{spt} + T_{a,frc}\varepsilon_{frc1} \qquad (6.9)$$

$$\hat{V}_w = V_{w,0} + V_{w,spt}\varepsilon_{spt} + V_{w,frc}\varepsilon_{frc2}$$

$$\hat{\phi} = \phi_0 + \phi_{spt}\varepsilon_{spt} + \phi_{frc}\varepsilon_{frc3}$$

$$\hat{I} = I_0 + I_{frc}\varepsilon_{frc4}$$

By plugging these affine forms into the AA-based counterpart of (6.4), (6.7), and (6.8), it is possible to compute the affine forms of the convective, radiated and solar flux, namely $\hat{q}_c$, $\hat{q}_r$, and $\hat{q}_s$, as far as the affine form of the conductor resistance $\hat{R}$. Consequently, the affine form of the static thermal rating can be determined as follows:

$$\hat{I}_{max} = \sqrt{\frac{\hat{q}_c + \hat{q}_r - \hat{q}_s}{\hat{R}}} \tag{6.10}$$

The solution of this AA-based equation returns the central value and the partial deviations of the static thermal rating, namely

$$\hat{I}_{max} = I_0 + I_{age}\varepsilon_{age} + I_{pol}\varepsilon_{pol} + \sum_{i=1}^{3} I_{spt,i}\varepsilon_{spt,i} + \sum_{i=1}^{3} I_{frc,i}\varepsilon_{frc,i} + I_{n+1}\varepsilon_{n+1} \tag{6.11}$$

where ε_{n+1} is a new noise symbol describing the cumulative effect of the endogenous uncertainties induced by the approximation errors of the nonaffine operators.

By using the same approach, it is possible to determine the affine form describing the time evolution of the conductor temperature in the presence of a step load variation by solving the following affine differential equation:

$$\frac{d\hat{T}_c}{dt} = \frac{1}{mC_p}\left[\hat{q}_s + \hat{I}^2\hat{R} - \hat{q}_c - \hat{q}_r\right] \tag{6.12}$$

which can be solved using an Euler-based method, hence, obtaining the following affine form for each discrete time step t_k:

$$\hat{T}_c(t_k) = T_{c,0}(t_k) + T_{c,age}(t_k)\varepsilon_{age} + T_{c,pol}(t_k)\varepsilon_{pol} + \sum_{i=1}^{3} T_{c,spt,i}(t_k)\varepsilon_{spt,i}$$

$$+ \sum_{i=1}^{3} T_{c,frc,i}(t_k)\varepsilon_{frc,i} + T_{c,n+1}(t_k)\varepsilon_{n+1} \tag{6.13}$$

The affine forms reported in (6.11) and (6.13) makes it possible to assess the impacts of each source of input data uncertainty on the static and transient thermal ratings.

6.1.3 Application Examples

The previously described AA-based method has been applied to calculate the thermal ratings of a 795 kcmil 26/7 ACSR conductor in the presence of data uncertainties (Piccolo et al., 2004). For this purpose, the perturbing effects of aging and pollutant emissions on the thermal model parameters and the forecasting errors and spatial drift of the weather variables have been assumed as the main sources of uncertainty.

Hence, each uncertain input variable is modeled by an affine form, which is composed of a central value, corresponding to its "nominal" value, and three partial deviations describing how the considered independent sources of uncertainty affect the variable value.

In particular, since the uncertain parameters α and τ vary in the range [0.23, 0.98] depending on the conductor aging and pollutant emissions, and considering the scarcity of experimental data (which makes it impossible to separately estimate the impact of each uncertainty source on the parameter values), a single noise symbol has been considered when defining the affine forms for these parameters.

Moreover, since the main uncertainty source affecting the total heat flux Q_s is related to the atmospheric pollutant levels, a single-noise symbol has been considered when defining the affine form $\hat{Q}_s$. The corresponding parameters of this affine form can be determined by computing the values of Q_s for both clear and industrial atmospheres, which represent the two extreme operating conditions. The corresponding results define the range of Q_s and the associated affine form.

As far as the weather variables are concerned, they are perturbed by two sources of uncertainty; these are related to spatial drift and forecasting errors. Assumptions regarding these sources of uncertainty have been posited in light of the experimental studies described in Piccolo et al. (2004).

Several numerical simulations have been performed to compute the static and dynamical thermal ratings for the overhead line, whose main characteristic data have been summarized in Table 6.1.

As far as the time evolution of the weather variables is concerned, the profiles reported in Figures 6.1 and 6.2 have been considered for the wind speed, wind direction, environmental temperature, and solar heat flux. In particular, the wind speed and direction profiles reported in Figure 6.1 have been considered as the central values of the affine forms $\hat{V}_w$ and $\hat{\phi}$, while the corresponding partial deviations have been determined by assuming tolerances of $\pm5\%$ and $\pm10\%$ for the spatial

Table 6.1 Simulation data.

Parameter name	Value
Simulated time	09:00 a.m.–14:00 p.m.
Latitude	40.85°N
Maximum conductor temperature	90 °C
Resistance for $T_c = 25$ °C	7.2835×10^{-5} Ω/m
Resistance for $T_c = 75$ °C	8.6877×10^{-5} Ω/m
Line orientation	East–west direction

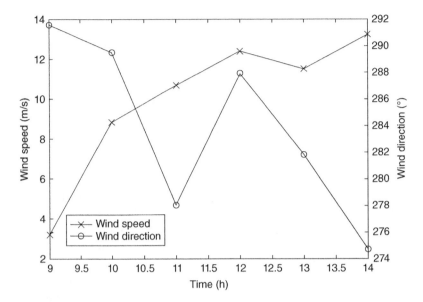

Figure 6.1 Deterministic profiles of the wind speed and direction.

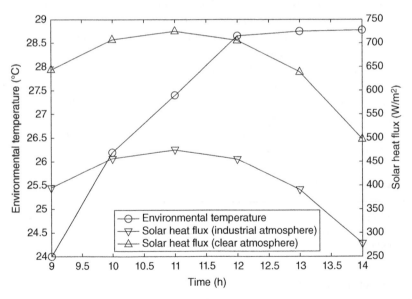

Figure 6.2 Deterministic profiles of the environmental temperature and solar heat flux for clear and industrial atmosphere.

drift and the forecasting errors, respectively. The same approach has been adopted when defining the affine form $\hat{T}_a$; in this case, the tolerances representing the spatial drift and the forecasting errors have been assumed as $\pm 3\%$ and $\pm 5\%$, respectively. Finally, the affine form modeling the solar heat flux has been defined by computing the central value and the radius of the range defined by the two extreme profiles reported in Figure 6.2. In this case, the effects of the forecasting errors and the spatial drift have been neglected.

The static thermal rating computed by applying the AA-based approach is reported in Figure 6.3. In this figure, the obtained results have been compared with those obtained by applying interval analysis (IA)-based computing and a Monte Carlo-based method.

The analysis of the obtained results confirms the effectiveness of both AA and IA-based computing to reliably identify the bounds of the static thermal ratings. Also in this application, AA-based computing can assess the impact of each source of input data uncertainty on the thermal rating solution, which is important and useful information.

The benefits of deploying AA-based computing have also been confirmed in dynamic thermal rating calculations. For this purpose, the differential equation

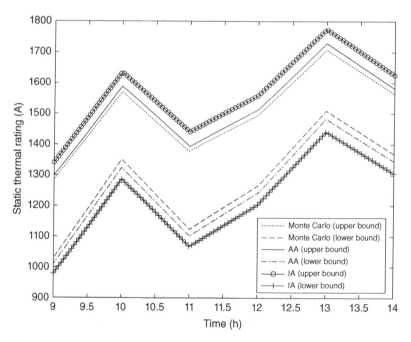

Figure 6.3 Bounds of the static thermal ratings computed by AA, IA, and Monte Carlo-based method.

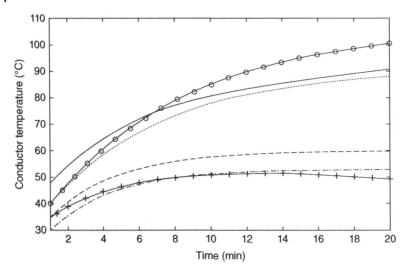

Figure 6.4 Bounds of the conductor temperature for a step load change of 900 A computed by AA, IA, and Monte Carlo-based method.

describing the conductor heat exchange for a step load change of 900 A has been solved by considering the effects of the same data uncertainties considered before. The time evolution of the conductor temperature profile computed by AA, IA, and Monte Carlo-based method is reported in Figure 6.4.

The analysis of the obtained results confirms the effectiveness of both AA and IA to compute reliable enclosures of the dynamic trajectories of the conductor temperature for all the possible combinations of the input data uncertainties.

6.2 Thermal Rating Assessment of Power Cables

Power cables thermal analysis is a crucial issue for ensuring the reliable operation of modern power systems, as these cables are frequently the limiting factor of the overall transfer capability. They could also play a significant role in enhancing the flexibility of transmission and distribution networks. Determining the exact thermal condition of cables is essential for various power system control tasks, such as load management, emergency grid operation, and contingency management.

Specifically, thermal analysis of power cables uses reliable mathematical models to estimate the time-evolution of the hot spot conductor temperature and identify the maximum allowable loading levels that the line can tolerate based on of its current thermal state and expected environmental conditions. While many models for power cables thermal analysis have been proposed in literature, their deployment can be hindered by uncertainties in both the model parameters and input variables (Hanna et al., 1998).

Thermal modeling of power cables is subject to uncertainty from various sources, such as randomness in weather data, inaccuracies in load predictions, and imprecise estimations of the cable thermal parameters. These sources of uncertainty can significantly impact the results of thermal analysis. Hence, it is essential to perform a thorough tolerance analysis to model the effects of data uncertainty and integrate these into the thermal modeling process, as referenced in Villacci and Vaccaro (2007).

Conducting a systematic tolerance analysis enables the evaluation of the uncertainty in input data and the propagation of that uncertainty to the output of the thermal model. This provides a better understanding of the level of confidence when estimating the thermal state of power cables. Additionally, it can effectively support comprehensive assessment of the thermal model sensitivity to large changes in the input data, allowing for an estimation of how the output of the thermal model will change in response to input variations.

IA-based methods could prove to be an invaluable solution to this problem, as highlighted in sources (Villacci and Vaccaro, 2007; Götz and Günter, 2000).

6.2.1 Thermal Modeling of Power Cables

The mathematical model describing the power cable thermal dynamics can be formalized by discretizing the insulation and the jacket regions in n_i and n_p isothermal coaxial cylindrical layers, respectively (Götz and Günter 2000). The thermal dynamics of each cell located between two consecutive layers can be modeled using the equivalent circuit shown in Figure 6.5.

where R_j, C_j^{int}, and C_j^{ext} are the thermal resistance and the internal and external thermal capacity of the jth layer, respectively. These thermal parameters can be

Figure 6.5 Thermal modeling of the jth cell.

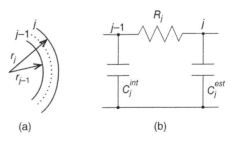

(a)   (b)

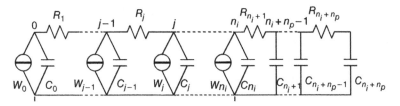

Figure 6.6 Equivalent thermal model of the power cables.

determined as follows:

$$R_j = \frac{\rho}{2\pi} \ln \frac{r_j}{r_{j-1}}$$

$$C_j^{int} = \pi c_p \left[\left(\frac{r_j + r_{j-1}}{2} \right)^2 - r_{j-1}^2 \right]$$

$$C_j^{ext} = \pi c_p \left[r_j^2 - \left(\frac{r_j + r_{j-1}}{2} \right)^2 \right]$$

(6.14)

where r_j and r_{j-1} are the radius of the jth and $(j-1)$th layer, respectively, while ρ is the thermal resistivity and c_p is the specific heat. The combination of the thermal circuits of all $n_i + n_p$ cells produces the equivalent thermal model of the power cables; the schematic of which is reported in Figure 6.6.

Where the generation parameters W_j aim at modeling the internal heating generation phenomena, e.g. ohmic and dielectric losses, and the thermal capacitance can be determined as

$$C_0 = C_c + C_1^{int}$$

$$C_j = C_j^{ext} + C_{j+1}^{int} \quad \forall j \in [1, n_i - 1]$$

$$C_{n_i} = C_{n_i}^{ext} + C_g + C_{n_i+1}^{int}$$

$$C_{n_i+n_p} = C_{n_p}^{ext}$$

(6.15)

The dynamic behavior of this equivalent thermal circuit is described by the following set of ordinary differential equations:

$$C_0 \frac{dT_0}{dt} = W_0 - \frac{T_0 - T_1}{R_1}$$

$$C_1 \frac{dT_1}{dt} = \frac{T_0 - T_1}{R_1} - \frac{T_1 - T_2}{R_2}$$

(6.16)

$$\cdots$$

$$C_{n_i+n_p-1} \frac{dT_{n_i+n_p-1}}{dt} = \frac{T_{n_i+n_p-2} - T_{n_i+n_p-1}}{R_{n_i+n_p-1}} - \frac{T_{n_i+n_p-1} - T_{n_i+n_p}}{R_{n_i+n_p-1}}$$

This set of equations can be represented by the following matrix formulation:

$$
\begin{bmatrix} \frac{dT_0}{dt} \\ \frac{dT_1}{dt} \\ \dots \\ \frac{dT_{n_i+n_p-1}}{dt} \end{bmatrix} = \mathbf{A} \begin{bmatrix} T_0 \\ T_1 \\ \dots \\ T_{n_i+n_p-1} \end{bmatrix} + \mathbf{B} \begin{bmatrix} W_0 \\ 0 \\ \dots \\ T_{n_i+n_p} \end{bmatrix} \tag{6.17}
$$

where the matrices $\mathbf{A}$ and $\mathbf{B}$ can be determined as follows:

$$
\mathbf{A} = \begin{bmatrix} -\frac{1}{R_1 C_0} & \frac{1}{R_1 C_0} & 0 & 0 & \dots & 0 & 0 \\ \frac{1}{R_1 C_1} & -\left(\frac{1}{R_1}+\frac{1}{R_2}\right)\frac{1}{C_1} & \frac{1}{R_2 C_1} & 0 & \dots & 0 & 0 \\ \dots & & & & & & \\ 0 & 0 & 0 & 0 & \dots & \frac{1}{R_{n_i+n_p-1} C_c} & -\left(\frac{1}{R_{n_i+n_p-1}}+\frac{1}{R_{n_i+n_p}}\right)\frac{1}{C_{n_i+n_p-1}} \end{bmatrix} \tag{6.18}
$$

$$
\mathbf{B} = \begin{bmatrix} \frac{1}{C_0} & 0 & \dots & 0 \\ 0 & 0 & \dots & 0 \\ \dots & & & \\ 0 & 0 & \dots & \frac{1}{R_{n_i+n_p} C_{n_i+n_p}} \end{bmatrix} \tag{6.19}
$$

The solution of this dynamic thermal model calculates the time evolution of the temperatures of the cable layers, thus, and the corresponding evolution of the hot spot temperature.

6.2.2 Sources of Data Uncertainties

The described thermal model is affected by large and heterogeneous uncertainties, such as fluctuations of the thermophysical soil properties, drift of the conductor thermal parameters, spatial distribution of the weather variables, and load forecasting errors. In particular, experimental studies have shown that the thermal resistivity and capacity of the materials used in power cables are subject to considerable variations during their lifetime. Additionally, there is a significant degree of uncertainty when determining soil thermal characteristics under different environmental and loading conditions; this is due to the complex relationship between soil type, density, and moisture content, as can be inferred by analyzing the tolerances of the main thermophysical soil parameters summarized in Table 6.2.

Table 6.2 Thermophysical soil parameters.

Soil	Thermal conductivity (W/m/K)	Density (kg/m^3)	Specific heat (kJ/kg/K)
Dry sandy soil	0.4–0.6	1400–1600	0.80–0.85
Sandy-gravelly soil	0.8–2.0	1600–2600	0.86–0.88
Vegetable soil	1.0–1.2	1400–1500	1.1–1.4
Clayey soil	1.2–1.8	2000–2500	0.80–0.90
Volcanic soil	0.9–1.6	1500–2300	0.70–0.80

When analyzing the reported data, it is important to note that variations in thermophysical properties can also be observed within the same soil type. This variability is primarily due to factors such as the size and shape of individual soil components, the degree of crystallization, and the characteristics of the porosity.

Also in this case, data uncertainty arises from the spatial variability and forecasting errors of weather variables.

6.2.3 Tolerance Analysis of Cable Thermal Dynamics by IA

Let us define the uncertain vector $\mathbf{p} = (p_1, \ldots, p_n)$, whose elements are the uncertain input parameters, and the corresponding known tolerance $[\mathbf{P}] = ([P_1], \ldots, [P_n])$. These uncertainties result in uncertain cable temperatures, which vary within their tolerances $[\mathbf{T}]$. Hence, the overall problem is to determine the time evolution of the cable temperature tolerance once the tolerance of the input data has been fixed by solving the following set of interval dynamic equations:

$$\frac{d\mathbf{T}}{dt} = \mathbf{A}(\mathbf{p})\mathbf{T} + \mathbf{B}(\mathbf{p})\mathbf{u}(\mathbf{p}) \quad \forall \mathbf{p} \in [\mathbf{P}] \tag{6.20}$$

where it has been assumed that both matrices $\mathbf{A}$ and $\mathbf{B}$ depend on the uncertain input vector.

Let us denote $T(t, \mathbf{p})$ as the solution of (6.20) for some fixed $\mathbf{p} \in [\mathbf{P}]$. Then the solution set of (6.20) for the time interval $[0, t_f]$ can be defined as

$$S(0, t_f) = \left\{ T(t, \mathbf{p}) : t \in [0, t_f], \mathbf{p} \in [\mathbf{P}] \right\} \tag{6.21}$$

while the following set:

$$S(t) = \{T(t, \mathbf{p}) : \mathbf{p} \in [\mathbf{P}]\} \tag{6.22}$$

is defined as the reachability set at time t, and the smallest set of interval vectors $\mathbf{X}(t)$ containing $S(t) \forall t \in [0, t_f]$ is defined as the interval solution of (6.20). A reliable approximation of this interval solution set can be computed by deploying an Euler-based solution scheme using conventional IA-based operators (Götz and Günter, 2000).

6.2.4 Application Examples

In this example, the described IA-based technique is applied to compute the bounds of the hot-spot temperature of the underground power cables depicted by the schematic in Figure 6.7. The cable characteristic data are summarized in Table 6.3. For the purpose of this example, the uncertainty of the model

Figure 6.7 Schematic of the analyzed power cable system.

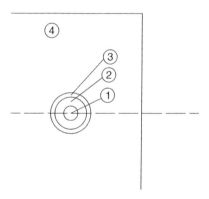

Table 6.3 Thermophysical properties of the power cable components.

Component	k (W/m/K)	ρ (kg/m³)	c (J/kg/K)	d (cm)
Conductor (1)	400	8.90×10^3	380	1.14
Insulating material (2)	0.22	1.10×10^3	1.75×10^3	2.54
Jacket material (3)	0.160	1.40×10^3	1.30×10^3	3.22

Table 6.4 Thermophysical properties of the soil.

Material	k (W/m/K)	ρ (kg/m³)	c (J/kg/K)
Humid sandy soil (4)	[0.4, 0.6]	[1900, 2000]	[860, 900]
Native soil (5)	[1.0, 1.4]	[1400, 1600]	[800, 850]

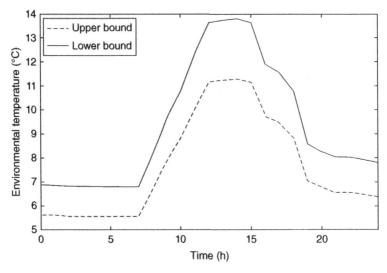

Figure 6.8 Time evolution of the environmental temperature.

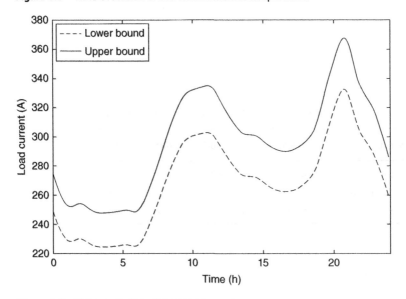

Figure 6.9 Time evolution of the load current.

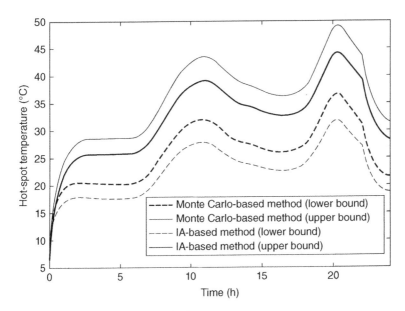

Figure 6.10 Time evolution of the hot-spot temperature.

parameters has been taken into account by considering the thermophysical variations of the materials surrounding the cables. The corresponding tolerances are reported in Table 6.4.

As far as the time evolution of the environmental temperature and load current are concerned, the profiles reported in Figures 6.8 and 6.9, which are characterized by a tolerance of ±10% and ±5%, respectively, have been assumed.

The corresponding bounds of the hot-spot cable temperature computed by the described IA-based technique and by a Monte Carlo-based method are reported in Figure 6.10.

From these results, it can be clearly seen that the tolerance solution obtained using the IA-based approach is more conservative than the solution obtained using the Monte Carlo method.

References

Hadi Ahmadi and Ahad Kazemi. The islanded micro-grid large signal stability analysis based on neuro-fuzzy model. *International Transactions on Electrical Energy Systems*, 30(8):e12449, 2020.

Morteza Aien, Ali Hajebrahimi, and Mahmud Fotuhi-Firuzabad. A comprehensive review on uncertainty modeling techniques in power system studies. *Renewable and Sustainable Energy Reviews*, 57:1077–1089, 2016.

Fernando Alvarado and Zian Wang. *Direct sparse interval hull computations for thin non-M matrices*. 1993.

Luciano V. Barboza, Graçaliz P. Dimuro, and Renata H.S. Reiser. Towards interval analysis of the load uncertainty in power electric systems. In *Proceedings of the International Conference on Probabilistic Methods Applied to Power Systems*, pages 538–544, 2004.

Yakov Ben-Haim and Isaac Elishakoff. *Convex Models of Uncertainty in Applied Mechanics*. Elsevier, 2013.

W.Z. Black, S.S. Collins, and J.F. Hall. Theoretical model for temperature gradients within bare overhead conductors. *IEEE Transactions on Power Delivery*, 3(2):707–715, 1988. doi: 10.1109/61.4309.

Hans Blanc and Dick Den Hertog. On Markov chains with uncertain data. 2008.

Kevin Brokish and James Kirtley. Pitfalls of modeling wind power using Markov chains. *2009 IEEE/PES Power Systems Conference and Exposition*, pages 1–6, 2009.

Ennio Brugnetti, Guido Coletta, Fabrizio De Caro, Alfredo Vaccaro, and Domenico Villacci. Enabling methodologies for predictive power system resilience analysis in the presence of extreme wind gusts. *Energies*, 13(13):3501, 2020.

S.Q. Bu, W. Du, and H.F. Wang. Investigation on probabilistic small-signal stability of power systems as affected by offshore wind generation. *IEEE Transactions on Power Systems*, 30(5):2479–2486, 2014.

Robert Calvin Burchett and G.T. Heydt. Probabilistic methods for power system dynamic stability studies. *IEEE Transactions on Power Apparatus and Systems*, PAS-97(3):695–702, 1978.

Interval Methods for Uncertain Power System Analysis, First Edition. Alfredo Vaccaro.
© 2023 The Institute of Electrical and Electronics Engineers, Inc. Published 2023 by John Wiley & Sons, Inc.

R.C. Burchett, H.H. Happ, and K.A. Wirgau. Large scale optimal power flow. *IEEE Transactions on Power Apparatus and Systems*, PAS-101(10):3722–3732, 1982.

Claudio Canizares, T. Fernandes, E. Geraldi Jr., L. Gérin-Lajoie, M. Gibbard, I. Hiskens, J. Kersulis, R. Kuiava, L. Lima, F. Marco, et al. Benchmark systems for small signal stability analysis and control. https://resourcecenter.ieee-pes.org/publications/technical-reports/PESTR18.html, (PES-TR), 2015.

Snehashish Chakraverty and Saudamini Rout. Affine arithmetic based solution of uncertain static and dynamic problems. *Synthesis Lectures on Mathematics and Statistics*, 12(1):1–170, 2020.

Richard D. Christie. Power systems test case archive. In *available on line at* https://labs.ece.uw.edu/pstca/, 1999.

Assem Deif. The interval eigenvalue problem. *ZAMM-Journal of Applied Mathematics and Mechanics/Zeitschrift für Angewandte Mathematik und Mechanik*, 71(1):61–64, 1991.

Elizabeth D. Dolan and Jorge J. Moré. Benchmarking optimization software with performance profiles. *Mathematical Programming*, 91(2):201–213, 2002.

D.A. Douglass. Weather-dependent versus static thermal line ratings (power overhead lines). *IEEE Transactions on Power Delivery*, 3(2):742–753, 1988. doi: 10.1109/61.4313.

Stewart N. Ethier and Thomas G. Kurtz. *Markov Processes: Characterization and Convergence*, volume 282. John Wiley & Sons, 2009.

Giorgio M. Giannuzzi, Viktoriya Mostova, Cosimo Pisani, Salvatore Tessitore, and Alfredo Vaccaro. Enabling technologies for enhancing power system stability in the presence of converter-interfaced generators. *Energies*, 15(21):8064, 2022.

Antonio Gomez-Exposito, Antonio J. Conejo, and Claudio A. Ca nizares. *Electric Energy Systems: Analysis and Operation*. CRC Press, 2009.

Alicia Mateo González, A.M. Son Roque, and Javier García-González. Modeling and forecasting electricity prices with input/output hidden Markov models. *IEEE Transactions on Power Systems*, 20(1):13–24, 2005.

Darius Grabowski, Markus Olbrich, and Erich Barke. Analog circuit simulation using range arithmetics. In *Proceedings of the Asia and South Pacific Design Automation Conference*, pages 762–767, 2008.

J. Grainger and W. Stevenson. *Power System Analysis*. McGraw-Hills, 1994.

Samuele Grillo, Antonio Pievatolo, and Enrico Tironi. Optimal storage scheduling using Markov decision processes. *IEEE Transactions on Sustainable Energy*, 7(2):755–764, 2015.

Alefeld Götz and Mayer Günter. Interval analysis: theory and applications. *Journal of Computational and Applied Mathematics*, 121(1–2):421–464, 2000.

C. Ryan Gwaltney, Youdong Lin, Luke D. Simoni, and Mark A. Stadtherr. Interval methods for nonlinear equation solving applications. In *Handbook of Granular Computing*. John Wiley & Sons, 2008.

M.A. Hanna, A.Y. Chikhani, and M.M.A. Salama. Thermal analysis of power cable systems in a trench in multi-layered soil. *IEEE Transactions on Power Delivery*, 13(2):304–309, 1998. doi: 10.1109/61.660894.

Ian A. Hiskens and Jassim Alseddiqui. Sensitivity, approximation, and uncertainty in power system dynamic simulation. *IEEE Transactions on Power Systems*, 21(4):1808–1820, 2006.

Milan Hladík. Optimal value bounds in nonlinear programming with interval data. *Top*, 19(1):93–106, 2011.

Youqin Huang. An interval algorithm for uncertain dynamic stability analysis. *Applied Mathematics and Computation*, 338:567–587, 2018.

Huazhang Huang, C.Y. Chung, Ka Wing Chan, and Haoyong Chen. Quasi-Monte Carlo based probabilistic small signal stability analysis for power systems with plug-in electric vehicle and wind power integration. *IEEE Transactions on Power Systems*, 28(3):3335–3343, 2013.

IEEE Std 738-1993. IEEE standard for calculating the current-temperature of bare overhead conductors. pages 1–48, 1993. doi: 10.1109/IEEESTD.1993.120365.

J.P. Jans. On Frobenius algebras. *Annals of Mathematics*, 69(2):392–407, 1959.

Katy Klauenberg and Clemens Elster. Markov chain Monte Carlo methods: An introductory example. *Metrologia*, 53(1):S32, 2016.

L. Kolev. A general interval method for global nonlinear DC analysis. In *Proceedings of the European Conference on Circuit theory and design*, volume 97, pages 1460–1462, 1997.

L. Kolev and I. Nenov. A combined interval method for global solution of nonlinear systems. In *Proceedings of the XXIII International Conference on Fundamentals of Electronics and Circuit Theory*, pages 365–368, 2000.

Igor O. Kozine and Lev V. Utkin. Interval-valued finite Markov chains. *Reliable Computing*, 8(2):97–113, 2002.

Prabha Kundur, Neal J. Balu, and Mark G. Lauby. *Power System Stability and Control*, volume 7. McGraw-hill New York, 1994.

Prabha Kundur, John Paserba, Venkat Ajjarapu, Göran Andersson, Anjan Bose, Claudio Canizares, Nikos Hatziargyriou, David Hill, Alex Stankovic, Carson Taylor, et al. Definition and classification of power system stability IEEE/CIGRE joint task force on stability terms and definitions. *IEEE Transactions on Power Systems*, 19(3):1387–1401, 2004.

Andreas Lemke, Lars Hedrich, and Erich Barke. Analog circuit sizing based on formal methods using affine arithmetic. In *Proceedings of the IEEE/ACM International Conference on Computer-Aided Design*, pages 486–489, 2002.

V.I. Levin. Nonlinear optimization under interval uncertainty. *Cybernetics and Systems Analysis*, 35(2):297–306, 1999.

V.I. Levin. Comparison of interval numbers and optimization of interval-parameter systems. *Automation and Remote Control*, 65(4):625–633, 2004.

Xuexin Liu, Wai-Shing Luk, Yu Song, Pushan Tang, and Xuan Zeng. Robust analog circuit sizing using ellipsoid method and affine arithmetic. In *Proceedings of the 2007 Asia and South Pacific Design Automation Conference*, pages 203–208, 2007.

Xindong Liu, Mohammad Shahidehpour, Zuyi Li, Xuan Liu, Yijia Cao, and Zhaohong Bie. Microgrids for enhancing the power grid resilience in extreme conditions. *IEEE Transactions on Smart Grid*, 8(2):589–597, 2016.

V. Loia and A. Vaccaro. Decentralized economic dispatch in smart grids by self-organizing dynamic agents. *IEEE Transactions on Systems, Man, and Cybernetics: Systems*, 44(4):397–408, 2014.

David G. Luenberger. Introduction to dynamic systems: Theory, models, and applications. Technical report, 1979.

Federico Milano, Florian Dörfler, Gabriela Hug, David J. Hill, and Gregor Verbič. Foundations and challenges of low-inertia systems. In *2018 Power Systems Computation Conference (PSCC)*, pages 1–25. IEEE, 2018.

James A. Momoh. A generalized quadratic-based model for optimal power flow. In *Proceedings of the IEEE International Conference on Systems, Man and Cybernetics*, pages 261–271, 1989.

James A. Momoh and J.Z. Zhu. Improved interior point method for OPF problems. *IEEE Transactions on Power Systems*, 14(3):1114–1120, 1999.

Ramon E. Moore. *Interval Analysis*. SIAM, 1966.

Giuseppe Muscolino and Alba Sofi. Stochastic analysis of structures with uncertain-but-bounded parameters via improved interval analysis. *Probabilistic Engineering Mechanics*, 28:152–163, 2012.

Werner Oettli and William Prager. Compatibility of approximate solution of linear equations with given error bounds for coefficients and right-hand sides. *Numerische Mathematik*, 6(1):405–409, 1964.

Yanfei Pan, Feng Liu, Laijun Chen, Jianhui Wang, Feng Qiu, Chen Shen, and Shengwei Mei. Towards the robust small-signal stability region of power systems under perturbations such as uncertain and volatile wind generation. *IEEE Transactions on Power Systems*, 33(2):1790–1799, 2017.

K.S. Pandya and S.K. Joshi. A survey of optimal power flow methods. *Journal of Theoretical & Applied Information Technology*, 4(5):450–458, 2008.

María José Pardo and David de la Fuente. Fuzzy Markovian decision processes: Application to queueing systems. *Computers & Mathematics with Applications*, 60(9):2526–2535, 2010.

Parikshit Pareek and Hung D. Nguyen. Probabilistic robust small-signal stability framework using Gaussian process learning. *Electric Power Systems Research*, 188:106545, 2020.

Carl E. Pearson. *Handbook of Applied Mathematics: Selected Results and Methods*, Springer. 1990.

Antonio Pepiciello and Alfredo Vaccaro. Small-signal stability analysis of uncertain power systems via interval analysis. *Electric Power Systems Research*, 212:108339, 2022.

Antonio Pepiciello, Fabrizio De Caro, Alfredo Vaccaro, and Sasa Djokic. Affine arithmetic-based reliable estimation of transition state boundaries for uncertain Markov chains. *Electric Power Systems Research*, 204:1–11, 2022.

Antonio Piccolo, Alfredo Vaccaro, and Domenico Villacci. Thermal rating assessment of overhead lines by Affine Arithmetic. *Electric Power Systems Research*, 71(3):275–283, 2004.

Mehrdad Pirnia, Claudio A. Ca nizares, and Kankar Bhattacharya. Revisiting the power flow problem based on a mixed complementarity formulation approach. *IET Generation, Transmission & Distribution*, 7(11):1194–1201, 2013.

Robin Preece and Jovica V. Milanović. Assessing the applicability of uncertainty importance measures for power system studies. *IEEE Transactions on Power Systems*, 31(3):2076–2084, 2015.

Robin Preece, Nick C. Woolley, and Jovica V. Milanović. The probabilistic collocation method for power-system damping and voltage collapse studies in the presence of uncertainties. *IEEE Transactions on Power Systems*, 28(3):2253–2262, 2012.

Robin Preece, Kaijia Huang, and Jovica V. Milanović. Probabilistic small-disturbance stability assessment of uncertain power systems using efficient estimation methods. *IEEE Transactions on Power systems*, 29(5):2509–2517, 2014.

Zhiping Qiu, Xiaojun Wang, and Michael I. Friswell. Eigenvalue bounds of structures with uncertain-but-bounded parameters. *Journal of Sound and Vibration*, 282(1–2):297–312, 2005.

J.L. Reding. A method for determining probability based allowable current ratings for BPA's transmission lines. *IEEE Transactions on Power Delivery*, 9(1):153–161, 1994. doi: 10.1109/61.277689.

Graham Rogers. *Power System Oscillations*. Springer Science & Business Media, 2012.

Jiri Rohn. Systems of linear interval equations. *Linear Algebra and its Applications*, 126:39–78, 1989. ISSN 0024-3795.

José L Rueda, Delia G. Colome, and Istvan Erlich. Assessment and enhancement of small signal stability considering uncertainties. *IEEE Transactions on Power Systems*, 24(1):198–207, 2009.

S.M. Rump. Accurate solution of dense linear systems, Part II: Algorithms using directed rounding. *Journal of Computational and Applied Mathematics*, 242:185–212, 2013.

Peter W. Sauer, Mangalore A. Pai, and Joe H. Chow. *Power System Dynamics and Stability: With Synchrophasor Measurement and Power System Toolbox*. John Wiley & Sons, 2017.

Koushik Sen, Mahesh Viswanathan, and Gul Agha. Model-checking Markov chains in the presence of uncertainties. In *International Conference on Tools and*

Algorithms for the Construction and Analysis of Systems, pages 394–410. Springer, 2006.

Jae Woong Shim, Gregor Verbič, and Kyeon Hur. Stochastic eigen-analysis of electric power system with high renewable penetration: Impact of changing inertia on oscillatory modes. *IEEE Transactions on Power Systems*, 35(6):4655–4665, 2020.

Raymond R. Shoults and D.T. Sun. Optimal power flow based upon PQ decomposition. *IEEE Transactions on Power Apparatus and Systems*, (2):397–405, 1982.

Philippe Smets. Imperfect information: Imprecision and uncertainty. In *Uncertainty Management in Information Systems*, pages 225–254. Springer, 1997.

Junseok Song, Vaidyanathan Krishnamurthy, Alexis Kwasinski, and Ratnesh Sharma. Development of a Markov-chain-based energy storage model for power supply availability assessment of photovoltaic generation plants. *IEEE Transactions on Sustainable Energy*, 4(2):491–500, 2012.

Bruce Stephen, Xiaoqing Tang, Poppy R. Harvey, Stuart Galloway, and Kyle I. Jennett. Incorporating practice theory in sub-profile models for short term aggregated residential load forecasting. *IEEE Transactions on Smart Grid*, 8(4):1591–1598, 2015.

Jorge Stolfi and Luiz Henrique De Figueiredo. Self-validated numerical methods and applications. In *Proceedings of the Monograph for 21st Brazilian Mathematics Colloquium*. Citeseer, 1997.

Study Committee 23-CIGRE. Dynamic loading of transmission equipment. *Electra*, 202, 202, 2002.

David I. Sun, Bruce Ashley, Brian Brewer, Art Hughes, and William F. Tinney. Optimal power flow by Newton approach. *IEEE Transactions on Power Apparatus and Systems*, PAS-103(10):2864–2880, 1984.

Giorgio Tognola and Rainer Bacher. Unlimited point algorithm for OPF problems. *IEEE Transactions on Power Systems*, 14(3):1046–1054, 1999.

Alfredo Vaccaro and Claudio A. Ca nizares. An affine arithmetic-based framework for uncertain power flow and optimal power flow studies. *IEEE Transactions on Power Systems*, 32:274–288, 2017.

Alfredo Vaccaro, Claudio A. Ca nizares, and Domenico Villacci. An affine arithmetic-based methodology for reliable power flow analysis in the presence of data uncertainty. *IEEE Transactions on Power Systems*, 25(2):624–632, 2010.

Alfredo Vaccaro, Claudio A. Ca nizares, and Kankar Bhattacharya. A range arithmetic-based optimization model for power flow analysis under interval uncertainty. *IEEE Transactions on Power Systems*, 28(2):1179–1186, 2013.

G. Verbic and C.A. Ca nizares. Probabilistic optimal power flow in electricity markets based on a two-point estimate method. *IEEE Transactions on Power Systems*, 21(4):1883–1893, 2006.

Domenico Villacci and Alfredo Vaccaro. Transient tolerance analysis of power cables thermal dynamic by interval mathematic. *Electric Power Systems Research*, 77(3):308–314, 2007.

Yih Huei Wan and Brian K. Parsons. *Factors Relevant to Utility Integration of Intermittent Renewable Technologies*. National Renewable Energy Laboratory, 1993.

Zhuoyao Wang, Mahshid Rahnamay-Naeini, Joana M. Abreu, Rezoan A. Shuvro, Pankaz Das, Andrea A. Mammoli, Nasir Ghani, and Majeed M. Hayat. Impacts of operators' behavior on reliability of power grids during cascading failures. *IEEE Transactions on Power Systems*, 33(6):6013–6024, 2018.

Zhao Xu, Z.Y. Dong, and P. Zhang. Probabilistic small signal analysis using Monte Carlo simulation. In *IEEE Power Engineering Society General Meeting, 2005*, pages 1658–1664. IEEE, 2005.

Zhao Xu, Mohsin Ali, Z.Y. Dong, and X. Li. A novel grid computing approach for probabilistic small signal analysis. In *2006 IEEE Power Engineering Society General Meeting*, pages 8–pp. IEEE, 2006.

Jinzhou Zhu and Yan Zhang. A frequency and duration method for adequacy assessment of generation systems with wind farms. *IEEE Transactions on Power Systems*, 34(2):1151–1160, 2018.

Chenxi Zhu, Yan Zhang, Zheng Yan, and Jinzhou Zhu. Markov chain-based wind power time series modelling method considering the influence of the state duration on the state transition probability. *IET Renewable Power Generation*, 13(12):2051–2061, 2019.

Ray Daniel Zimmerman, Carlos Edmundo Murillo-Sánchez, and Robert John Thomas. MATPOWER: Steady-state operations, planning, and analysis tools for power systems research and education. *IEEE Transactions on Power Systems*, 26(1):12–19, 2010.

Index

Interval Methods for Uncertain Power System Analysis, First Edition. Alfredo Vaccaro.
© 2023 The Institute of Electrical and Electronics Engineers, Inc. Published 2023 by John Wiley & Sons, Inc.

IEEE Press Series on Power and Energy Systems

Series Editor: Ganesh Kumar Venayagamoorthy, Clemson University, Clemson, South Carolina, USA.

The mission of the IEEE Press Series on Power and Energy Systems is to publish leading-edge books that cover a broad spectrum of current and forward-looking technologies in the fast-moving area of power and energy systems including smart grid, renewable energy systems, electric vehicles and related areas. Our target audience includes power and energy systems professionals from academia, industry and government who are interested in enhancing their knowledge and perspectives in their areas of interest.

Printed and bound by CPI Group (UK) Ltd, Croydon, CR0 4YY

16/04/2025

14658583-0004